STRUCTURAL ANALYSIS THROUGH SHORT QUESTIONS AND ANSWERS

STRUCTURAL ANALYSIS THROUGH SHORT QUESTIONS AND ANSWERS

By

Dr. R. Vaidyanathan
B.E.(Hons), M.Sc.(Engg), Ph. D., F.I.E., M.I.S.T.E.
Formerly Principal Government College of Engineering
Tirunelveli-627007
Tamil Nadu

Dr. P. Perumal
B.E.(Distn), M.E.(Stru), Ph. D., F.I.E., M.I.S.T.E.
(Formerly Professor and Head Department of Civil Engineering,
Government College of Engineering
Salem-636011)
Principal Sri Ranganathar Institute of Engineering and Technology
Athipalayam Coimbatore-641110,
Tamil Nadu

UNIVERSITY SCIENCE PRESS
(An Imprint of Laxmi Publications Pvt. Ltd.)
An ISO 9001:2008 Company
BENGALURU • CHENNAI • COCHIN • GUWAHATI • HYDERABAD
JALANDHAR • KOLKATA • LUCKNOW • MUMBAI • RANCHI • NEW DELHI
BOSTON (USA) • ACCRA (GHANA) • NAIROBI (KENYA)

STRUCTURAL ANALYSIS THROUGH SHORT QUESTIONS AND ANSWERS

Printed and bound in India
Typeset at Goswami Associates, Delhi
First Edition : 2004; Third Edition : 2016
ISBN 978-93-80386-50-8

Published in India by

UNIVERSITY SCIENCE PRESS
(An Imprint of Laxmi Publications Pvt. Ltd.)

An ISO 9001:2008 Company
113, GOLDEN HOUSE, DARYAGANJ,
NEW DELHI - 110002, INDIA
Telephone : 91-11-4353 2500, 4353 2501
Fax : 91-11-2325 2572, 4353 2528
www.laxmipublications.com info@laxmipublications.com

Branches

Bengaluru	080-26 75 69 30	
Chennai	044-24 34 47 26,	24 35 95 07
Cochin	0484-237 70 04,	405 13 03
Guwahati	0361-254 36 69,	251 38 81
Hyderabad	040-27 55 53 83,	27 55 53 93
Jalandhar	0181-222 12 72	
Kolkata	033-22 27 43 84	
Lucknow	0522-220 99 16	
Mumbai	022-24 91 54 15,	24 92 78 69
Ranchi	0651-220 44 64	

C—

CONTENTS

Preface to the Third Edition

It is really heartening to find, from the Publishers, the good reception to the first two editions of this book, throughout India.

In this revised edition, many more new questions have been included in various chapters as suggested by the Publishers. This book can be treated as an addendum to Structural Analysis volumes I and II by the same authors.

Besides, this volume can be immensely useful for candidates preparing for competitive examinations.

The authors will be happy to receive suggestions for the further improvement of this book.

—Authors

Preface to the First Edition

This slender volume is an offspring of the two volumes of **"Comprehensive Structural Analysis"** which are fast becoming popular among students and teachers.

When these two volumes are expanded to include more chapters such as "Finite Element Analysis", "Kani's Method", etc., these questions and answers would be an integral part of them.

The questions are graded from the easy to the stiff ones. Care has been taken to include "good" questions from several university papers, questions that probe the mind rather than the memory.

A mastery of this volume will prepare the students not only for their degree but also for GATE, CES and job interviews.

Let this little book help you delve into the intricacies of structures.

—Authors

ACKNOWLEDGEMENTS

We are not sure we saw farther than others, but we had a lot of giants on whose slippery shoulders we could stand to peek into the intricacies of Structural Analysis. We thank them all.

First and foremost, we acknowledge the following authors whose books enlightened us in several aspects.

Punmia, Jain and Jain, Gupta Pandit and Gupta, Prakash Rao, Ramamrutham, Reddy, Khurmi, Negi and Jangid, Vazirani and Ratwani, Ramchandra, Junnarkar and Shah and Kinney

Mr. M. Karthikai Pandian of Mepco Schlenk Engineering College, Sivakasi was patient enough to unscramble our maze of numbers and sketches and gave them font and figures. We are really indebted to him.

M/s Laxmi Publications Pvt. Ltd. have been really very quick, understanding and accommodative in bringing out this book early.

We both are greatly beholden to Seetha and Saraswathi, our wives and our family for putting up with our night owl habits and irregular mealtimes. They have suffered a lot and we are trying to prove they have not suffered in vain. We thank Dr. P. Manimegalai, for her help in sorting and writing of the University Questions chapterwise.

A lot of people gave us words of encouragement which sustained us in our grind.

We say, "thank you folks; the words may not have cost you much, but they were invaluable to us".

Several sets of students have served as sounding boards for us, the brilliant, the slow, the inattentive and the indifferent. But all of them have helped us decide what kind of presentation would be effective.

We therefore dedicate this book to the student community.

—Authors

1

Classification and Behaviour of Structures

1. Differentiate between elastic and inelastic behaviour of structures.

Ans. A structure is elastic, if upon removal of the loads from the structure, it returns to the predeformed position it had before the loads were applied.

A structure is inelastic, if upon removal of the loads the structure does not return to its original predeformed position but retains some deformation.

2. Distinguish between linear and non-linear behaviour of structures.

Ans. If the force-displacement relationship of a structure is linear, the behaviour of the structure is said to be linear. The behaviour is said to be non-linear if the force-displacement relationship is non-linear (see Fig. 1).

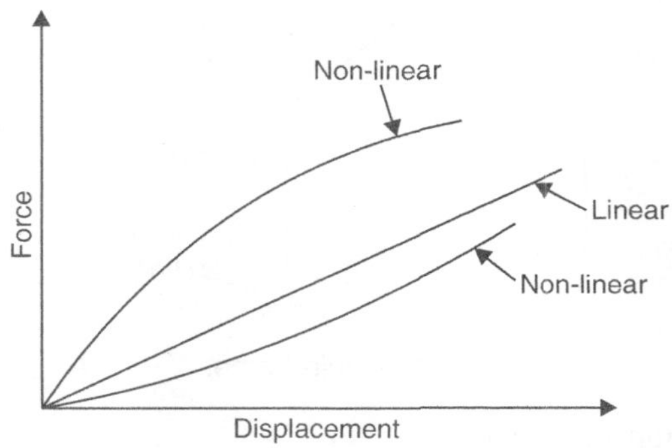

Fig. 1

3. What is meant by principle of superposition ?

Ans. Superposition of forces. The displacement at a section due to a number of loads acting simultaneously on a linearly elastic structure can be obtained by adding the displacements due to each load separately.

Superposition of displacements. The force developed at a section due to a number of displacements occurring simultaneously on a linearly elastic structure can be obtained by adding the force due to each displacement separately imposed.

Superposition of forces/displacements is permissible only in linear structures.

4. Distinguish between pin jointed and rigidly jointed structure.

Ans.

S.No.	*Pin Jointed Structure*	*Rigidly Jointed Structure*
1.	The joints permit change of angle between connected members.	The members connected at a rigid joint will maintain the angle between them even under deformation due to loads.
2.	The joints are incapable of transferring any moment to the connected members.	Members can transmit both forces and moments between themselves through the joint.
3.	The pins transmit forces between connected member by developing shear.	Provision of rigid joints normally increases the redundancy of the structures.

5. What are the assumptions made in the analysis of pin jointed trusses ?

Ans. Assumptions made in the analysis of pin jointed trusses are:

(*i*) External loads are transmitted to the structure only at the joint (or nodes).

(*ii*) Pins do not transfer any moment to any of the connected members.

(*iii*) Pins allow the connected members to change the angles between them.

6. What is a gable frame ?

Ans. A gable frame usually has two principal rafters inclined to the horizontal, to support a sloping roof. The frames can be pin jointed or rigidly jointed (see Fig. 2).

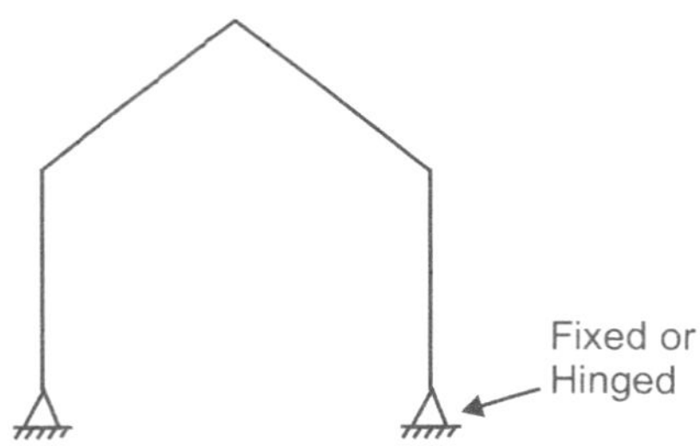

Fig. 2. Gable frame.

7. What is geometric non-linearity ?

Ans. Normally, we assume that deformations occurring in structures due to loads are very small. They are so small that they do not alter the geometry of the structure.

When deformations are large, they can change the geometry of a structure. Having caused a large deformation due to some loading we should realize that any further load acts on the altered structure. Thus the displacements will not be proportional to the loads. The load deformation relationship is thus non-linear. This non-linearity happening because of change of geometry is called geometric non-linearity.

8. How do material characteristics influence the behaviour of structures ?

Ans. Material properties like density, Young's modulus, Poisson's ratio, yield point, ductility do influence the behaviour of structures.

(*i*) Density determines the dead load.

(*ii*) Young's modulus determines the stiffness, deflection, buckling strength, ratio of sharing loads, etc.

(*iii*) Poisson's ratio does not have a great influence on structural behaviour except in composites, soils and behaviour of plates and shells.

(*iv*) Yield point is probably the most important attribute of a material and directly determines the load carrying capacity.

(*v*) Ductility becomes important when we deal with post yielding behaviour.

9. What is meant by lack of fit in a truss ?

Ans. One or more members in a pin jointed statically indeterminate frame may be a little shorter or longer than what is required. Such members will have to be forced in place during the assembling. These are called members having 'lack of fit'. Internal forces can develop in a redundant frame (without external loads) due to lack-of-fit.

10. A pin-jointed frame has 4 joints. How many members will be there, if it is indeterminate to one degree ?

Ans. Static indeterminacy, $I = 1$

No. of joints, $j = 4$

No. of members, $m = ?$

We know,

$$I = m - (2j - 3)$$

$$1 = m - (2 \times 4 - 3)$$

$$1 = m - 5$$

$$1 + 5 = m$$

$\therefore$ No. of members, $m = 6$.

11. Distinguish between a plane truss and a plane frame.

Ans. A plane truss is an articulated structure consisting members in a plane, pin jointed at their ends and carrying axial forces only.

A plane frame is a structure consisting members in a plane, rigid jointed at their ends and carrying bending moments, shear forces and axial forces.

The term, plane frame is sometimes loosely used to represent a plane truss also.

12. What are eigen value problems in structures ?

Ans. In eigen value problems, the stiffness matrix of a structure is liable to be singular. The determinant of the matrix would be zero. This would represent a state of instability. At this state the structure can switch its mode of taking external loads, for example, from compression mode to bending mode.

The structure may also keep performing periodic motions.

13. What is meant by thermal stresses ?

Ans. Thermal stresses are stresses developed in a structure/member due to change in temperature. Normally, determinate structures do not develop thermal stresses. They can absorb changes in lengths and consequent displacements without developing stresses.

14. Write down the two methods of determining displacements in pin jointed plane frames by the unit load concept.

Ans. The methods of using unit loads to compute displacements are,

(*i*) dummy unit load method

(*ii*) using the principle of virtual work.

15. The deflection at the free end of a cantilever due to the action of unit load at mid-span is $\frac{5k^3}{48EI}$. Then calculate the deflection at mid-span due to a unit load the free end. [Refer to Fig. 3 (*a*)]

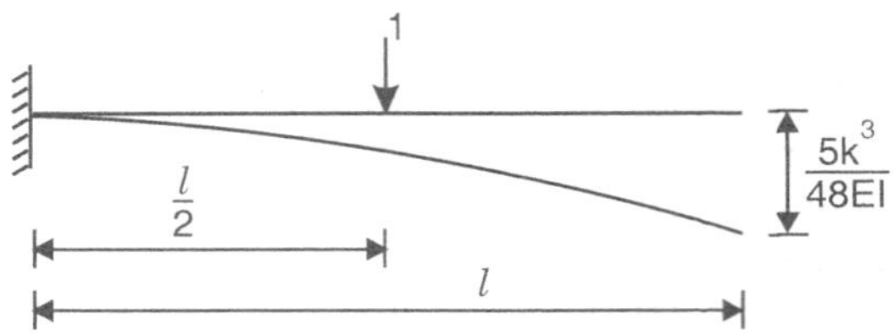

Fig. 3 (a)

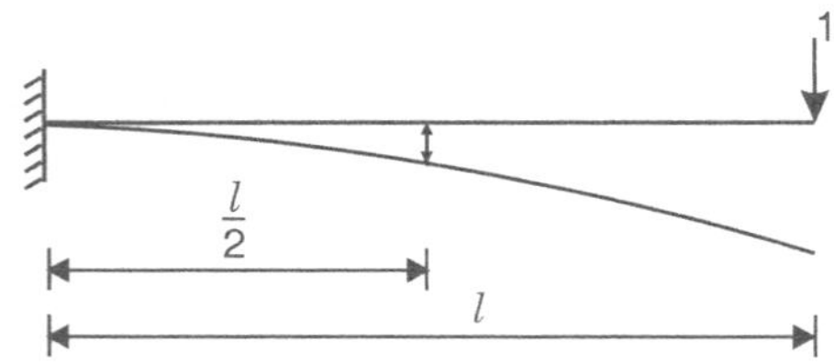

Fig. 3 (b)

Ans. The deflection at the mid span-due to unit load at the free end

$$= \frac{5k^3}{48EI} \text{ [vide Fig. 3 (}b\text{)]}$$

(By Maxwell's law of reciprocal deflection).

16. The prop reaction of a propped cantilever beam is 'R' due to certain loading. Write down the equation to solve for R.

Ans.

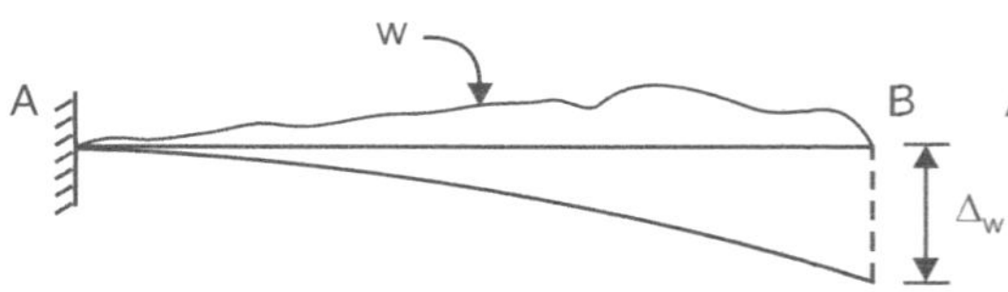

Fig. 4 (a)

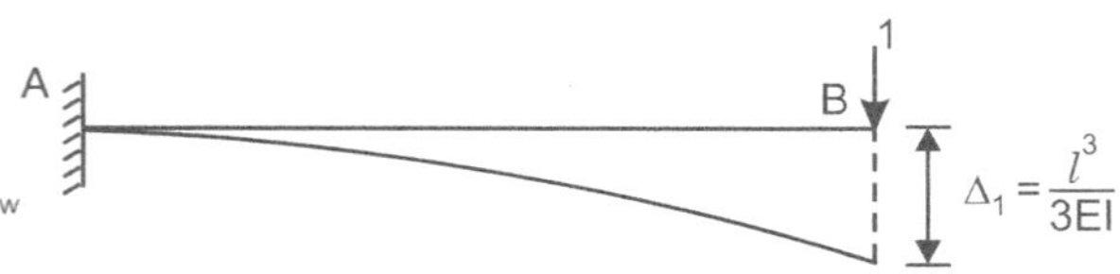

Fig. 4 (b)

Let Δ_w be the deflection at the free end B of the cantilever (with prop removed) due to the loading w. Let Δ_1 be the deflection due to a unit load at B. Then $R = \frac{\Delta_w}{\Delta_1} = \frac{\Delta_w}{l^3} 3EI$.

17. Mention the approximate methods of analysis of multi-storey frames.

Ans. Approximate methods of analysis of multi-storey frames are:

(*i*) Horizontal loads : (*a*) portal method

(*b*) cantilever method

(*c*) factor method.

(*ii*) Vertical loads : (*a*) point of inflection method

(*b*) substitute frame method.

18. What are the internal forces for which a beam member of a grid structure shall be designed ?

Ans. The members of a grid will have to be designed for

(*i*) bending moment

(*ii*) shear force and

(*iii*) twisting moment.

19. What is knee brace ?

Ans. Knee braces are provided at the beam column junctions of portals (see Fig. 5). They reduce the bending moments at the joints. Knee braces are mainly used in steel frames, especially in bridge portals.

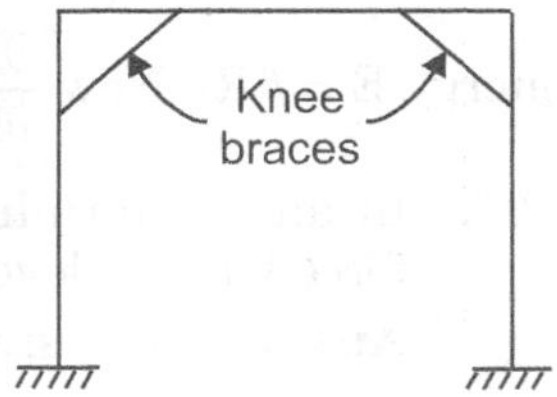

Fig. 5

20. What is the effect of temperature on the members of a statically determinate plane truss ?

Ans. In determinate structures temperature changes do not create any internal stresses. The changes in lengths of members may result in displacement of joints. But these would not result in internal stresses or changes in external reactions.

21. Distinguish between 'deck type' and 'through type' trusses.

Ans. A deck type truss is one in which the road is at the top chord level of the trusses. We would not see the trusses when we ride on the road way.

A through type truss is one in which the road is at the bottom chord level of the trusses. When we travel on the road way, we would see the web members of the trusses on our left and right. That gives us the impression that we are going 'through' the bridge.

22. What is a 'half through bridge' ?

Ans. A half through bridge is one in which the bridge deck is at mid height of the trusses. In these bridges, the cross girder loads would be transferred to the bottom chords like in through bridges.

23. Which method is preferable to analyse a grid structure ?

Ans. (*i*) In its simplest form, grids can be analysed as main girders and cross girders.

(*ii*) Grids may be analysed as orthotropic plates.

(*iii*) True analysis of grids can only be made by matrix (stiffness or flexibility) method. This alone can help us determine the shears, bending moments and torques in each member of the grid.

24. What is the advantage of adopting energy method in structural analysis ?

Ans. Energy methods are universal. Every structural system apportions forces in its members or subsystems so that the total energy is a minimum. If there are N forces (internal forces or reactions) every typical force F_i takes a value such that $\frac{\partial U}{\partial F_i} = 0$, where U is the total energy. Hence we get as many equations as there are forces to be determined.

The chief advantage is that F_i can be bending moment, torque, shear, thrust or support reaction. We can find them.

25. What is the strain energy stored in a rod of length l and axial rigidity AE due to an axial force P.

Ans. Strain energy stored, $U = \frac{P^2 l}{2AE}$

26. Give the principle of strain energy method to determine a redundant reaction.

Ans. In any structure, the internal forces and reactive forces take such values that will keep the strain energy to a minimum.

If there is an unknown redundant reaction R, let us get all the reactions and internal forces (like bending moments, shears and thrusts) in terms of R. Then we can get the strain energy $E = f(R)$. Now $\frac{\partial E}{\partial R} = 0$. Hence we can get R.

27. In an indeterminate truss, temperature of one member is increased. What nature of force will be developed in that member ?

Ans. Compressive force.

28. What is the use of counter diagonals in bridge trusses ?

Ans. In bridge trusses the diagonals of panels close to mid-span may develop either tension or compression depending upon the location of live loads (see Fig. 6).

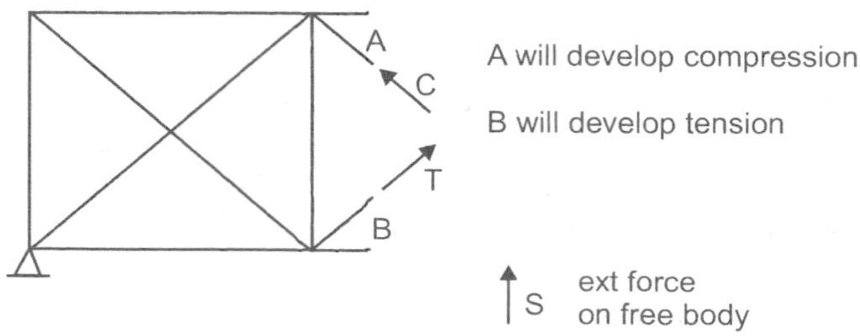

Fig. 6. Free body.

Compression members have to be designed against possible buckling. So they shall have a large moment of inertia, larger area and are more expensive.

It is cheaper to have 2 diagonals in the panel than one diagonal which can develop compression. When we have 2 diagonals, the connections are so made that the member becomes inactive in compression and active only in tension. This method of providing 2 braces in a panel is called counter bracing.

29. What is meant by 'reversal of stresses' in a member of a plane truss ?

Ans. In a bridge truss, diagonals of panels close to mid-span can develop tension for some live load positions and compression in other cases.

This causes reversal of stresses in those members (as live loads traverse the bridge).

30. In a two dimensional pin jointed truss, express the internal indeterminacy in terms of number of members and joints.

Ans. The internal indeterminacy of a two dimensional pin jointed truss is given by

$$I = m - (2j - 3)$$

where m = Number of members

j = Number of joints.

2

Static and Kinematic Indeterminacies of Structures

1. Define 'static indeterminacy' of structures.

Ans. If the conditions of statics *i.e.* $\Sigma H = 0, \Sigma V = 0$ and $\Sigma M = 0$ alone are not sufficient to find either external reactions or internal forces in a structure, the structure is called a statically indeterminate structure.

2. Define 'degree of redundancy' of a structure.

Ans. The degree of redundancy or indeterminacy is the number of unknown reaction components or stresses over and above the number of condition equations available for solution.

3. Classify statically redundant structures.

Ans. A statically redundant structure can be classified into two categories :

(*i*) **Externally redundant.** Structures having redundant reaction components.

(*ii*) **Internally redundant.** Structures having more members than those required for stability

A redundant structure may have external as well as internal redundancies.

4. Explain static indeterminacy with an example.

Ans. There are three conditions of static equilibrium, *i.e.* $\Sigma H = 0, \Sigma V = 0$ and $\Sigma M = 0$. If a structure can be analysed just by using these three conditions of equilibrium, then it is a statically determinate structure; otherwise it is called a statically indeterminate structure. The number of unknowns over and above the static condition equations available is called the static indeterminacy of the structure.

Example : Fixed beam with general loading

No. of unknowns = 6

No. of condition equations available = 3

$\therefore$ Static indeterminacy = 6 – 3 = 3.

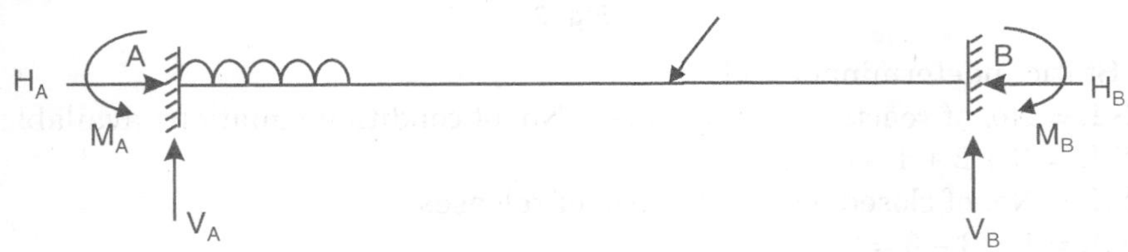

Fig. 1

5. Can a structure that is internally deficient be made stable with the use of external constraints ? Explain with an example.

Ans. Yes. In many cases. Consider the frame in Fig. 2(*a*). It is a pin-jointed deficient frame. For 4 joints we would need 4 × 2 – 3 = 5 members.

But it can be made stable by using 4 constraints instead of 3 as in Fig. 2(*b*).

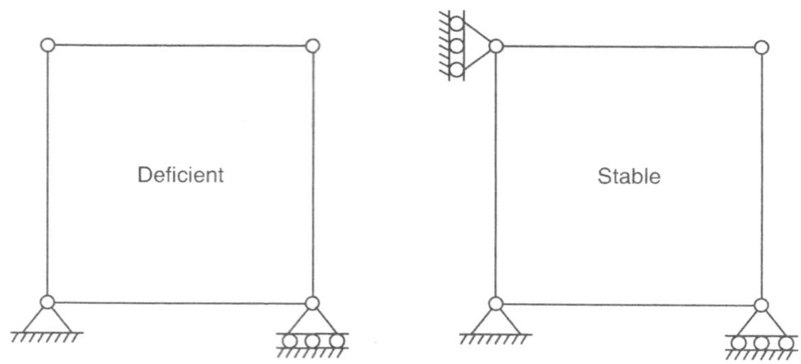

Fig. 2 (a) **Fig. 2** (b)

6. Find out the static indeterminacies of the following structures:

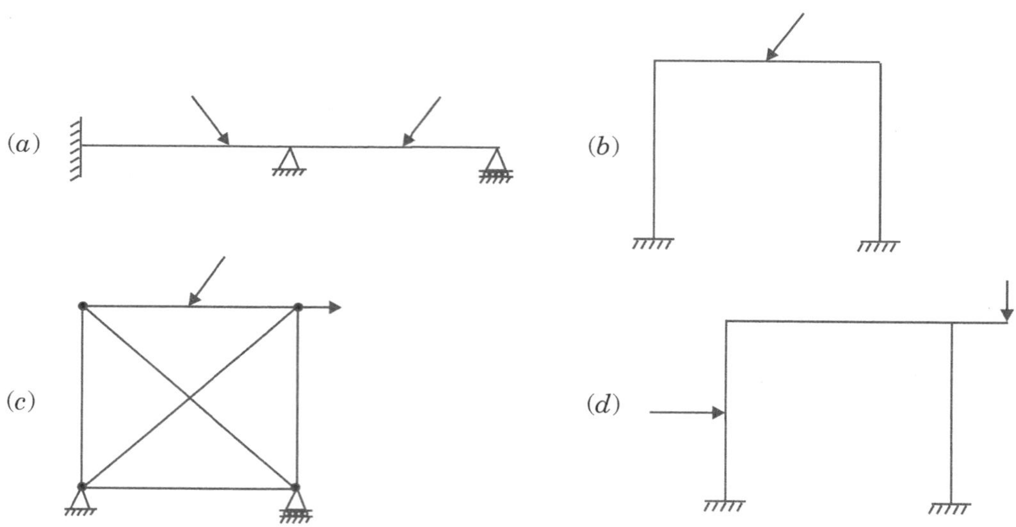

Fig. 3

Ans. Static indeterminacy (S.I) :

(*a*) S.I. = No. of reaction components – No. of condition equations available

S.I. = 3 + 2 + 1 – 3 = 3

(*b*) S.I. = No. of closed loops × 3 – No. of releases

S.I. = 1 × 3 – 0 = 3

(*c*) External indeterminacy = No. of reactions – No. of equilibrium equations

= 3 – 3 = 0

Internal indeterminacy $= m - (2j - 3)$
$= 6 - (2 \times 4 - 3) = 6 - 5 = 1$

Total indeterminacy = External + internal indeterminacy
$= 0 + 1 = 1$

(*d*) S.I. = No. of closed loops × 3 – No. of releases
$= 1 \times 3 - 0 = 3.$

7. Find the degree of static indeterminacy of the rigid jointed frame shown in Fig. 4.

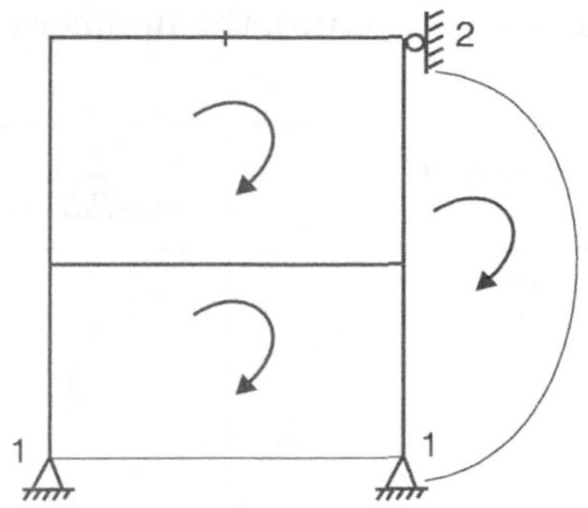

Fig. 4

Ans. S.I. = No. of closed loops × 3 – No. of releases
$= 3 \times 3 - 4 = 5.$

8. Give examples of beams of one degree static indeterminacy.
Ans. Beams with one degree static indeterminacy :

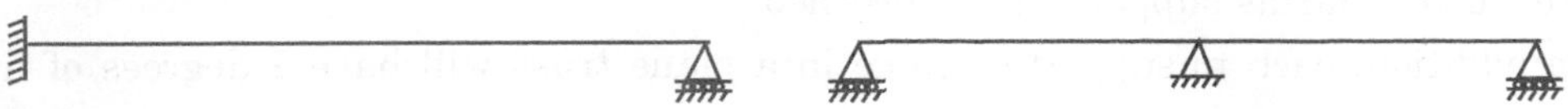

Fig. 5

9. What are the degrees of indeterminacy of a propped cantilever and a fixed beam?
Ans. Propped cantilever :
Static indeterminacy = 1
Kinematic indeterminacy = 2 (considering axial deformation also)

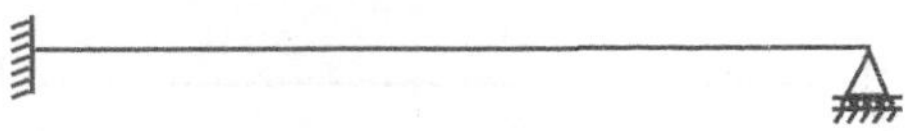

Fig. 6 (a)

Fixed beam :
Static indeterminacy = 2 (for vertical loading only)
= 3 (for general loading)
Kinematic indeterminacy = 0.

Fig. 6 (b)

10. Kinematically determinate structure need not be statically determinate – State True/False. Give an example.

Ans. True.

Example : Fixed beam in Fig. 6 (*b*).

11. How will you determine the kinematic indeterminacy of a beam structure ?

Ans. Kinematic indeterminacy refers to the number of possible joint displacements in a structure.

Example : In a two dimensional structure, the number of unknown joint displacements are as below :

Type of support	*Degrees of freedom (possible joint displacements)*
Roller 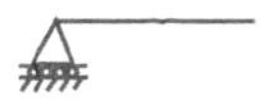	2
Hinges 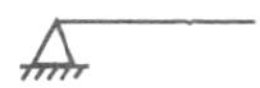	1
Fixity 	0

Depending upon the type and number of supports, the kinematic indeterminacy is obtained.

12. How do we assign kinematic indeterminacy of a plane truss structure ?

Ans. The definition of kinematic indeterminacy for a beam structure is valid for truss structures also as far as supports are concerned.

In addition, each unsupported hinge in a plane truss will have 2 degrees of freedom (K.I.).

13. Define 'kinematic indeterminacy' of a structure.

Ans. When a structure is subjected to loads, each joint will undergo displacements in the form of translations and rotations. Kinematic indeterminacy of a structure means the number of unknown joint displacements in a structure.

14. What is the kinematic indeterminacy of the beam shown in Fig. 7.

Fig. 7

Ans. Support A is fixed and hence can not have any displacement.

Support B is on rollers and hence have two degrees of freedom; *i.e.,* can have horizontal translation and rotation.

Thus the beam is kinematically indeterminate to the second degree.

In most practical cases, axial deformations being small can be neglected. In such a case, the beam would be analysed as a structure, kinematically indeterminate to the first degree.

15. Find out the kinematic indeterminacy of the following structures.
Ans.

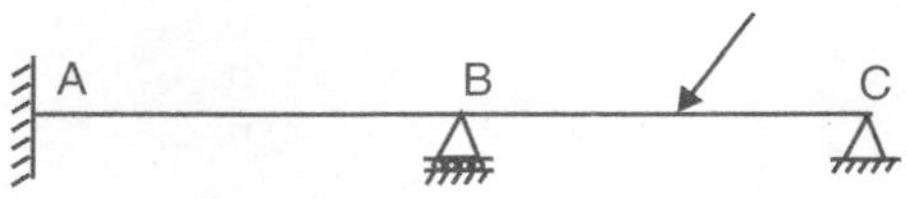

(a)

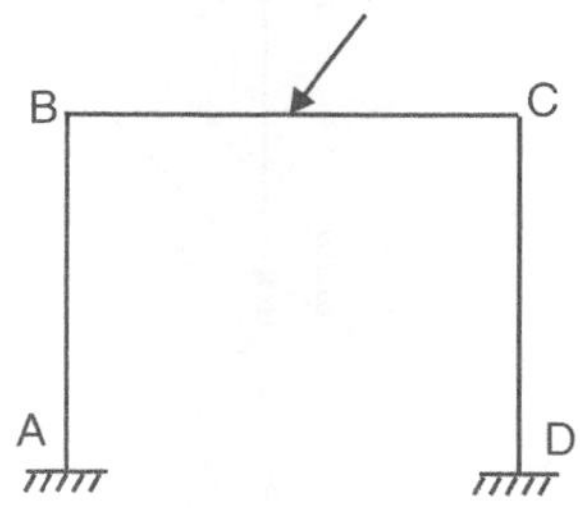

(b)

Fig. 8

For the continuous beam, K.I. = 2 ; (θ_B, θ_C)
For the portal frame, K.I. = 3 ; (θ_B, θ_C, Δ).

16. Find the degree of kinematic indeterminacy of the frame shown in Fig. 9. Neglect axial deformations in members.

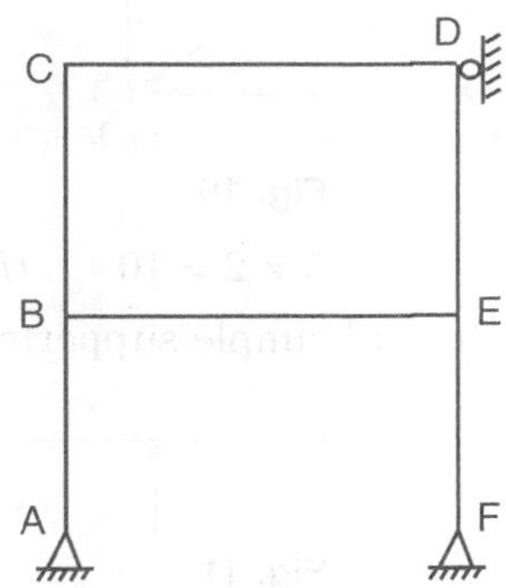

Fig. 9

Ans. K.I. = 7 ; (θ_A, θ_B, θ_C, θ_D, θ_E, θ_F, Δ_{BE}).

17. Find out the kinematic indeterminacy of the following structures (see Fig. 10).

(*i*)

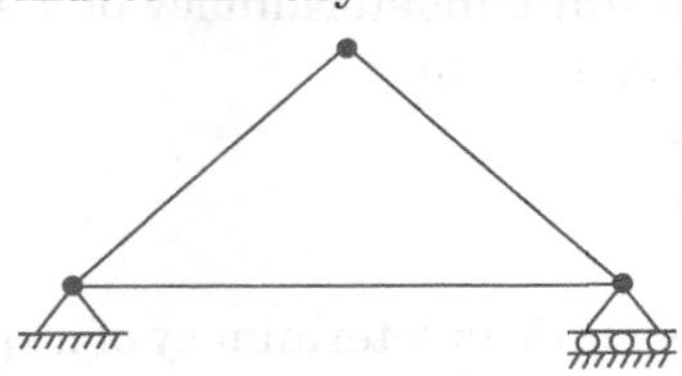

(*ii*)

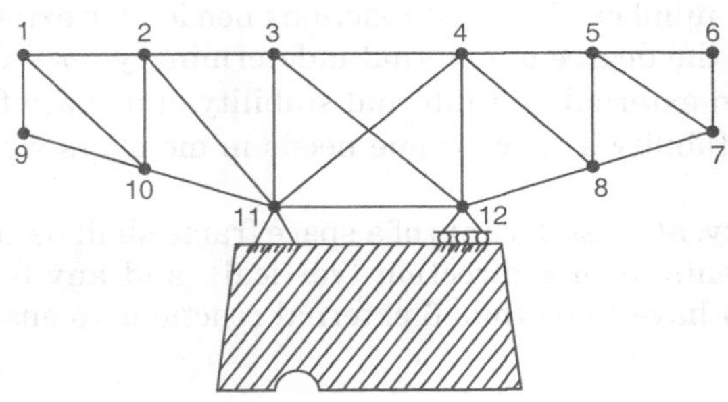

(*iii*)

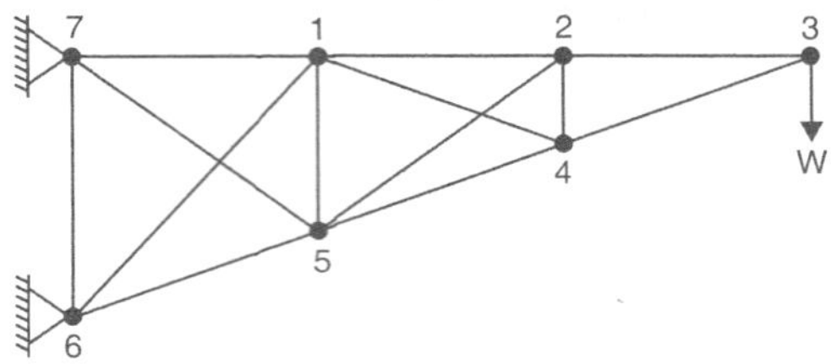

(*iv*)

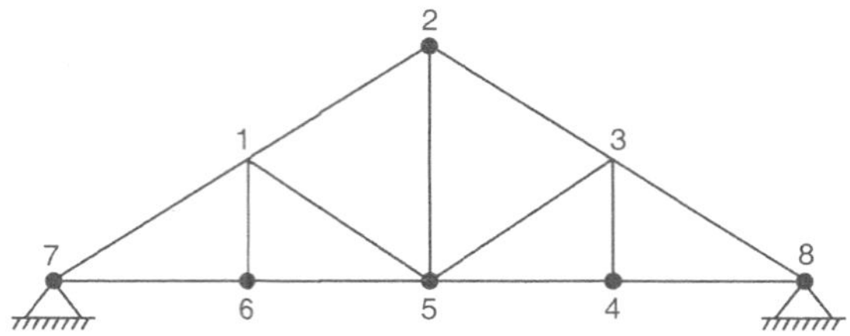

(*v*)

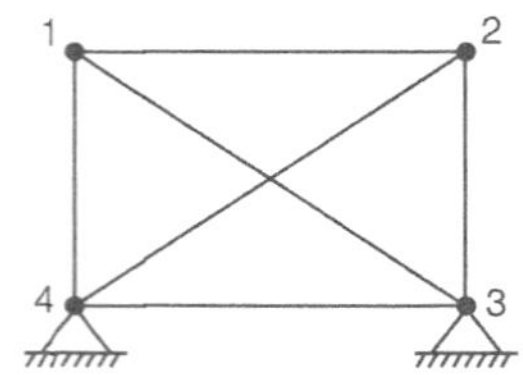

Fig. 10

Ans. (*i*) 3 (*ii*) 20 + 1 = 21 (*iii*) 5 × 2 = 10 (*iv*) 6 × 2 = 12 (*v*) 2 × 2 = 4.

18. A four span continuous beam has all simple supports. What is the static indeterminacy of this beam ?

Ans.

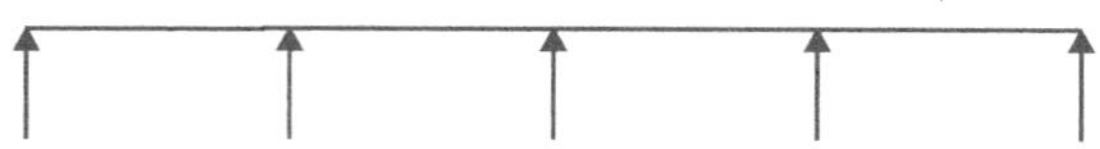

Fig. 11

Assuming the beam to be subjected to vertical loading only,

static indeterminacy = vertical reactions + end moments – 2

S.I. = 5 + 0 – 2 = 3.

19. Write the equation to find the static indeterminacy of a space truss.

Ans. For space trusses, S.I. = $m + r - 3j$

where, m = No. of members

r = No. of reactions

j = No. of joints.

20. Explain how external degree of static indeterminacy of a space truss can be determined ?

Ans. The minimum number of ground reactions needed for external stability is 6. If we have r external reactions, the degree of external indeterminacy = $r - 6$.

21. Distinguish between external and internal stability of a space frame structure.

Ans. For internal stability a space frame needs m members where $m = 3j - 6$, j being the number of joints.

For external stability, at least 3 points of a space frame shall be supported on the ground and 6 of their displacements (3 in z direction (vertical), and any 3 of 6 possible X and Y displacements). There will have to be thus 6 external reactions to ensure external stability.

22. A fixed beam is having an unsupported hinge at the mid-span. What is the kinematic indeterminacy of the beam ?

Ans.

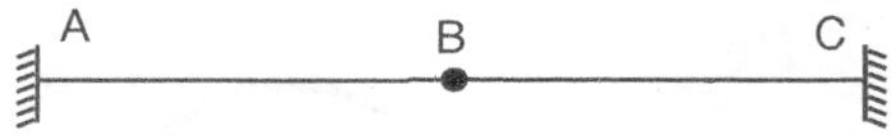

Fig. 12

Kinematic indeterminacy, K.I. = 2(Δ_B and θ_B).

23. Determine the degree of static indeterminacy of the beam shown in Fig. 13.

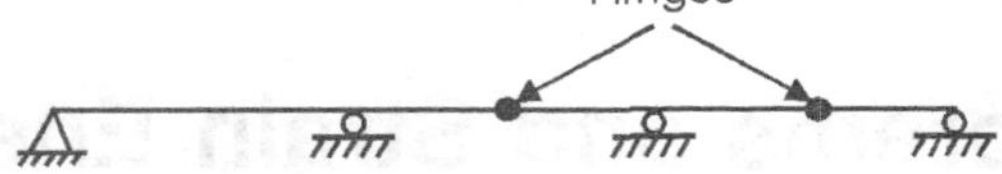

Fig. 13

Ans. Degree of static indeterminacy, I = (Vertical reactions + end moments – 2) – No. of internal hinges

$$= (4 + 0 - 2) - 2 = 0$$

∴ The given beam is statically determinate.

24. For the frame in Fig. 14, determine the static indeterminacy.

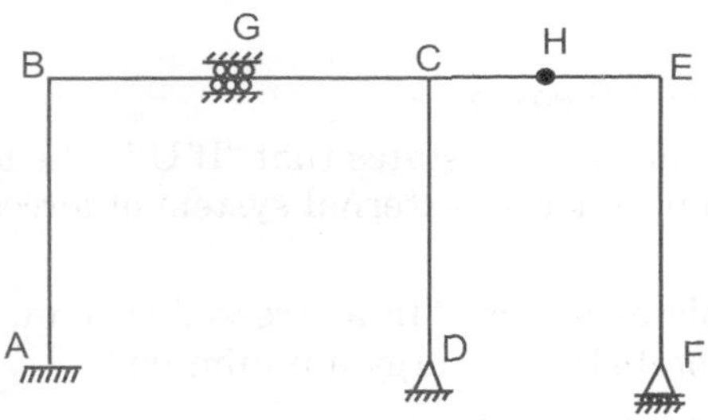

Fig. 14

Ans. No. of external restraints

at A : 3 at D : 2 at F : 1 at G : 2 Total : 8

No. of releases = 1 at H

No. of equilibrium equation = 3

S.I. = 8 – 1 – 3 = 4.

25. For the frame in Fig. 14, determine the K.I.

Ans. The degrees of freedom are marked in Fig. 15. This ignores axial deformations.

K.I. = 8

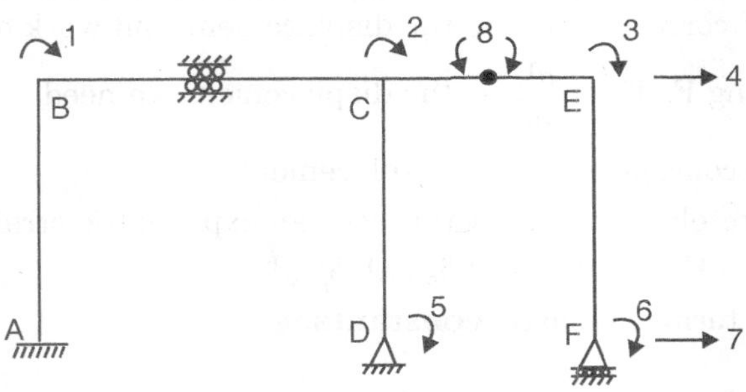

Fig. 15

3

General Theorems and Strain Energy Method

1. State Castigliano's first theorem.

Ans. Castigliano's first theorem states that the partial derivative of the total strain energy with respect to an applied force (or moment) gives the displacement (or rotation) at the point of application of the force (or moment) and in the direction of application of the force (or the moment).

2. State Castigliano's second theorem.

Ans. Castigliano's second theorem states that "If U is the total strain energy stored up in a frame work in equilibrium under an external system of forces, its magnitude is always a minimum".

We can also restate this as under : "In a stressed system, the (redundant) forces take such values as to make the total strain energy a minimum".

3. State the theorem useful for application to redundant structures.

Ans. The second theorem of Castigliano, is very useful in the application to redundant structures. This theorem is particularly useful for structures with members carrying axial forces, combined axial and bending forces, composite structures, etc. For each redundant force F_i we will get an equation based on the condition $\frac{\partial U}{\partial F_i} = 0$, where U is the total strain energy.

4. What is the use of Castigliano's first theorem ?

Ans. Castigliano's first theorem is used in the computation of deflections and rotations of structures. If we need the displacement at any point in a structure in a specified direction, we shall apply a force F corresponding to the displacement and work out the strain energy U due to all forces including F. Then $\frac{\partial U}{\partial F}$ is the displacement we need.

5. Substantiate the concept of virtual displacement.

Ans. In any discretely loaded structure we can express the strain energy U as

$$U = k\,[P_1\delta_1 + P_2\delta_2 \ldots P_i\delta_i \ldots] \qquad \ldots(3.1)$$

Here k will be in terms of elastic **constants.**

If we partially differentiate U with respect to any displacement δ_i, we get

$$\frac{\partial U}{\partial \delta_i} = k \cdot P_i$$

δ_i is called the virtual displacement and it helps us determine the force P_i.

6. What is the meaning and use of virtual forces ?

Ans. If we differentiate U [in equation (3.1)] with respect to P_i, we would get δ_i. Virtual force P_i helps us find δ_i.

7. A straight bar of length 300 mm has an uniform cross-sectional area of 200 mm^2. If an axial load 2.5 kN is applied, compute the strain energy stored in the bar if the Young's modulus of elasticity is 2×10^5 N/mm^2.

Ans. $l = 300$ mm ; A = 200 mm^2 ; P = 2.5 kN = 2500 N

$E = 2 \times 10^5$ N/mm^2

$$\text{Strain energy stored, } U = \frac{P^2 l}{2AE} = \frac{(2500)^2 \times 300}{2 \times 200 \times 2 \times 10^5} = 23.44 \text{ Nmm.}$$

8. A beam of span 5 m is subjected to a uniform bending moment of 400 kNm, applied gradually. Young's modulus of elasticity of the material is 2×10^5 N/mm^2 and the moment of inertia I = 450×10^6 mm^4. Calculate the strain energy stored in the beam.

Ans. $l = 5$ m ; M = 400 kNm

$E = 2 \times 10^5$ N/mm^2 = 2×10^8 kN/m^2 ; I = 450×10^6 mm^4 = 450×10^{-6} m^4

$EI = 2 \times 10^8 \times 450 \times 10^{-6} = 90,000$ kNm^2

$$\text{Strain energy stored, } U = \frac{M^2 l}{2EI} = \frac{(400)^2 \times 5}{2 \times 90000}$$

$$U = 4.444 \text{ kNm.}$$

9. Find the elongation of the bar in Fig. 1 using strain energy method.

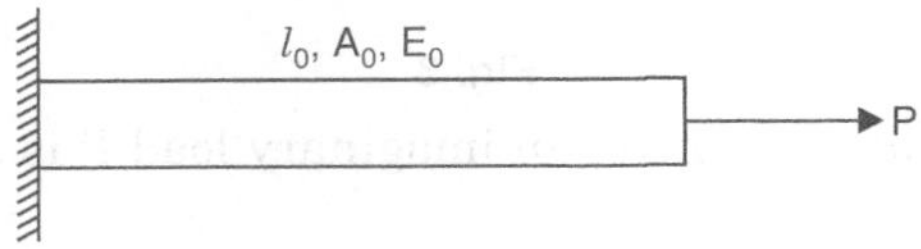

Fig. 1

Ans.

$$U = \frac{P^2 l}{2AE}$$

$$\delta = \frac{\partial U}{\partial P} = \frac{2Pl}{2AE} = \frac{Pl}{AE}$$

$$\delta = \frac{Pl}{AE}.$$

10. For the cantilever in Fig. 2 determine the slope at B

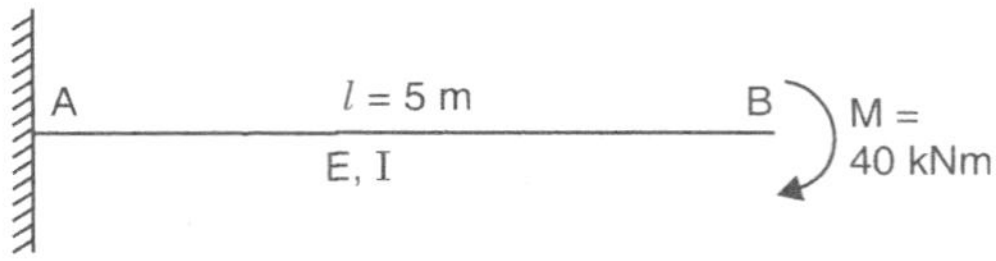

Fig. 2

using energy principles.

Ans. BM at any section at x from B is M

$$U = \int_0^l \frac{M^2 dx}{2EI}$$

$$\frac{\partial U}{\partial M} = \theta_B$$

Hence, $$\theta_B = \int_0^l \frac{2M\, dx}{2EI} = \frac{Ml}{EI}$$

$\therefore$ $$\theta_B = \frac{40 \times 5}{EI} = \frac{200}{EI} \text{ radian}$$

[E and I shall be in kN/m units]

11. A simply supported beam of span 4 m is subjected to a udl of 20 kN/m on its entire span. Write down the equation for finding the deflection at mid-span using strain energy method.

Ans.

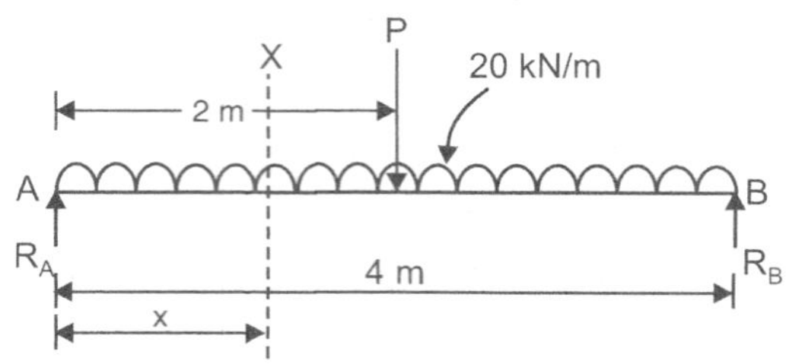

Fig. 3

To find the deflection at the centre, an imaginary load P is applied at the centre of the span

$$R_A = R_B = \frac{20 \times 4}{2} + \frac{P}{2} = 40 + \frac{P}{2}$$

$$M_x = \left(40 + \frac{P}{2}\right)x - \frac{20x^2}{2} = \left(40x + \frac{P}{2}x - 10x^2\right)$$

$$\frac{\partial M_x}{\partial P} = \frac{x}{2}$$

Equation for finding the deflection at mid-span.

$$\delta_c = \frac{1}{EI}\int_0^l M_x \frac{\delta M_x}{\delta P}\,dx$$

$$\delta_c = 2 \times \frac{1}{EI}\int_0^2 \left(40x + \frac{P}{2}x - 10x^2\right)\left(\frac{x}{2}\right)dx.$$

12. In the above problem, it is required to find the rotation of the left end A. Write down the equation to determine the rotation at A.

Ans.

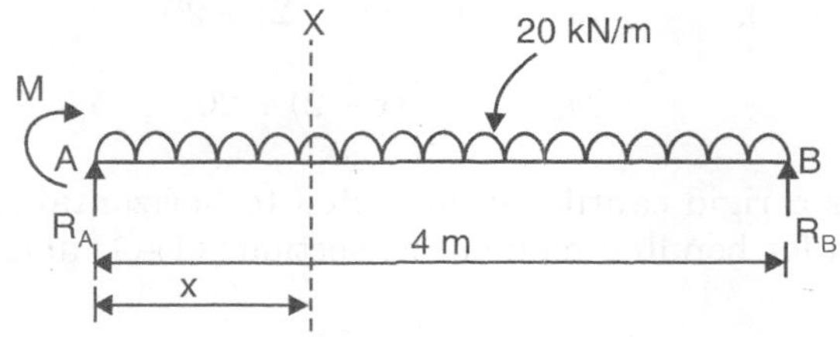

Fig. 4

To find θ_A, a fictitious moment M is applied at A.

$$R_A = \left(40 - \frac{M}{4}\right)$$

$$M_x = \left(40 - \frac{M}{4}\right)x + M - \frac{20x^2}{2}$$

$$M_x = 40x - \frac{M}{4}x + M - 10x^2$$

$$\frac{\partial M_x}{\partial M} = -\frac{x}{4} + 1$$

Equation for finding the rotation at A

$$\theta_A = \frac{\partial U}{\partial M} = \frac{1}{EI}\int_0^l M_x \frac{\partial M_x}{\partial M}\,dx$$

$$\theta_A = \frac{1}{EI}\int_0^4 \left(40x - \frac{M}{4}x + M - 10x^2\right)\left(-\frac{x}{4} + 1\right)dx.$$

13. For the beam shown in Fig. 5, write down the equations for bending moments at sections (1)-(1) and (2)-(2).

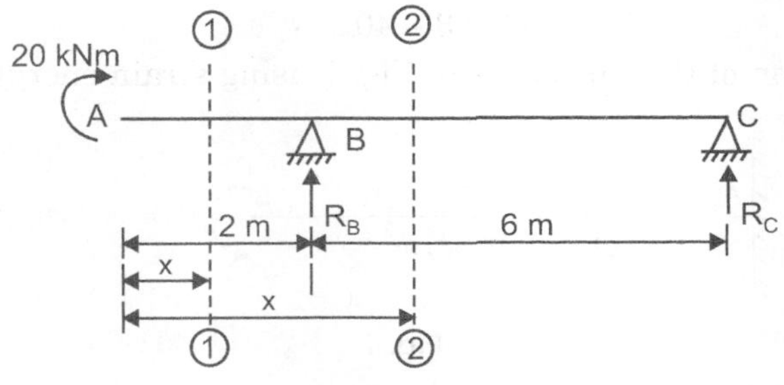

Fig. 5

Ans. Equating moments about C to zero,

$$R_B \times 6 + 20 = 0$$

$$R_B = -\frac{20}{6}$$

$$\therefore \quad R_C = \frac{20}{6}$$

At section (1)-(1), $M_x = 20$

At section (2)-(2), $M_x = -R_B \times (x - 2) + 20$

$$M_x = -\frac{20}{6}(x - 2) + 20$$

14. Fig. 6 shows a rigid cantilever subjected to horizontal and vertical loads. Write down the equation for bending moment at sections (1)-(1) and (2)-(2).

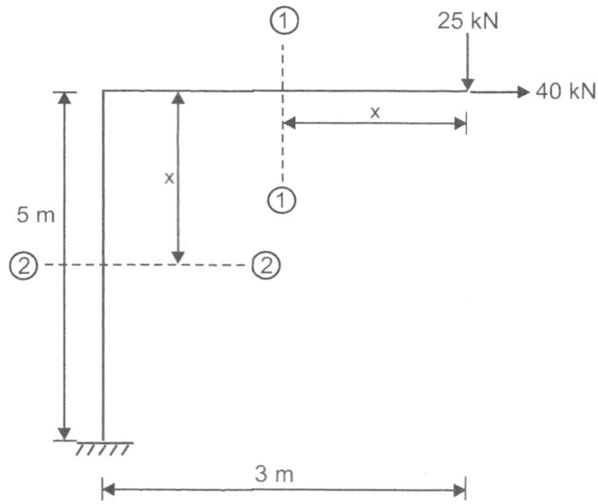

Fig. 6

Ans. Bending moment at section (1)-(1)

$$M_1 = -25x$$

Bending moment at section (2)-(2)

$$M_2 = -25 \times 3 - 40x.$$

15. Find the deflection of the cantilever in Fig. 7 using strain energy method.

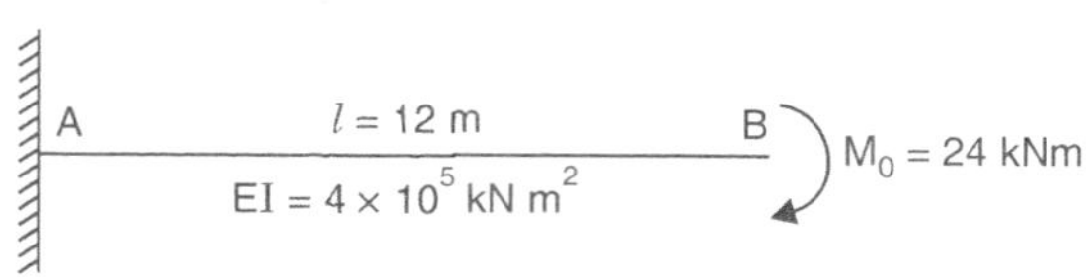

Fig. 7

Ans. We have to assume a virtual force P at B. [But P = O]

Now bending moment at any x from B is

$$M = M_0 + Px$$

$$U = \int_0^l \frac{M^2\,dx}{2EI}$$

$$\delta_B = \frac{\partial U}{\partial P} = \int \frac{2M \cdot \frac{\partial M}{\partial P}\,dx}{2EI} = \int_0^l \frac{(M_0 + Px)}{EI}(x)\,dx$$

$$\delta_B = \left(\frac{M_0 l^2}{2} + \frac{Pl^3}{3}\right) \Big/ EI$$

Since there is really no P,

$$\delta_B = \frac{M_0 l^2}{2EI} = \frac{24 \times 144}{8 \times 10^5} = 4.32 \times 10^{-3} \text{ m}$$

$$\delta_B = 4.32 \text{ mm.}$$

16. Find the displacement of C by Castigliano's theorem (see Fig. 8)

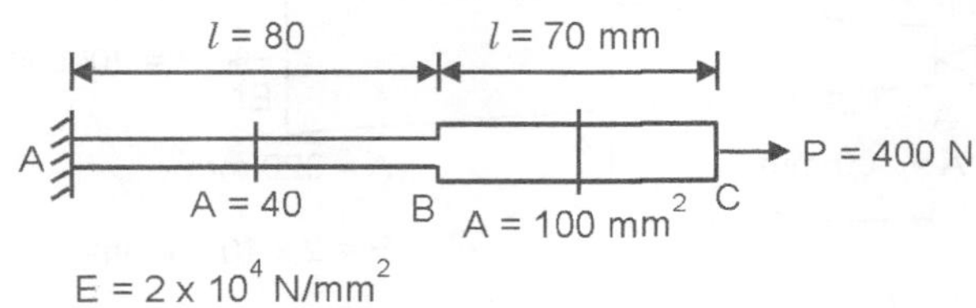

Fig. 8

Ans. Strain energy of the bar

$$U = U_{AB} + U_{BC}$$

$$= \Sigma \frac{P^2 l}{2AE}$$

$$U = \frac{P^2 \times 80}{80E} + \frac{P^2 \times 70}{200E} = \frac{P^2}{E}(1 + 0.35)$$

$$\delta_C = \frac{\partial U}{\partial P} = \frac{2P}{E}\,1.35 = \frac{2.7 \times 400}{2 \times 10^4} = 5.4 \times 10^{-2}$$

$$\delta_C = 0.054 \text{ mm.}$$

17. Explain Maxwell-Betti theorem taking the example of a cantilever and 2 points on the same.

Ans.

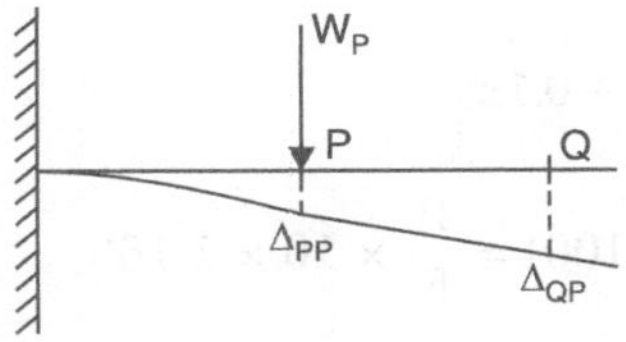

Fig. 9 (a). Load system A.

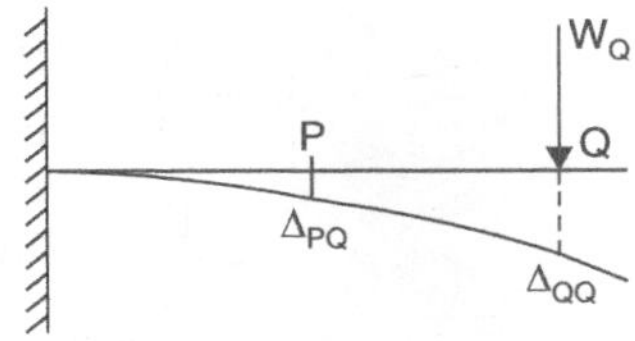

Fig. 9 (b). Load system B.

Suppose we have a structure on which either load system A or load system B can act.

According to Maxwell-Betti theorem, the work done by forces in system A as they move through the corresponding displacements in system B is equal to the work done by forces in system B as they move through the corresponding displacements in system A.

Accordingly $$W_P \times \Delta_{PQ} = W_Q \times \Delta_{QP} \quad ...(3.2)$$

18. From the results obtained from Maxwell-Betti theorem derive Maxwell's law of reciprocal deflection.

Ans. Maxwell's law says that deflection at A due to load at B is the same as the deflection at B due to the same load at A.

If we make $W_P = W_Q$ in equation (3.2), we get

$$\Delta_{PQ} = \Delta_{QP}$$

This is what Maxwell's law says.

19. Find the displacement of B by Castigliano's theorem. Sectional area varies linearly from A to B (see Fig. 10).

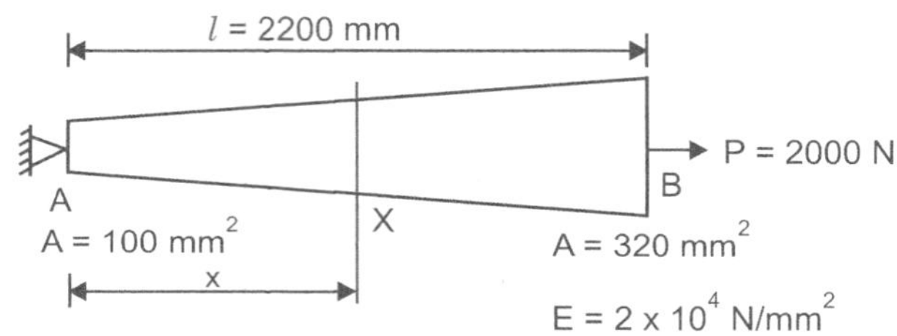

Fig. 10

Ans.
$$A_x = 100 + 220 \times \frac{x}{2200} = 100 + 0.1x$$

$$\sigma_x = \frac{P}{(100 + 0.1x)}$$

Strain energy,
$$U = \int_0^l \frac{\sigma_x^2}{2E} A_x \, dx$$

$$= \int \frac{P^2}{(100 + 0.1x)^2} (100 + 0.1x) \frac{dx}{2E}$$

$$= \frac{P^2}{2E} \int_0^l \frac{1}{100 + 0.1x} dx$$

$$U = \frac{P^2}{2E} \frac{1}{0.1} \Big[\log (100 + 0.1x) \Big]_0^l$$

$$\delta_B = \frac{\partial U}{\partial P} = \frac{P}{E} 10 \Big[\log (100 + 0.1x \Big]_0^l$$

$$= \frac{P}{E} 10 \, (\log 320 - \log 100) = \frac{P}{E} \times 10 \times 1.163$$

$$\delta_B = \frac{2000}{2 \times 10^4} \times 11.63 = 1.163 \text{ mm.}$$

20. A concrete beam has 10 per cent reinforcement.

$E_c = 2 \times 10^4$ $E_s = 2 \times 10^5$

Area of concrete (alone) = 900 mm^2

Area of steel = 100 mm^2

Work out the force taken by steel using Castigliano's second theorem.

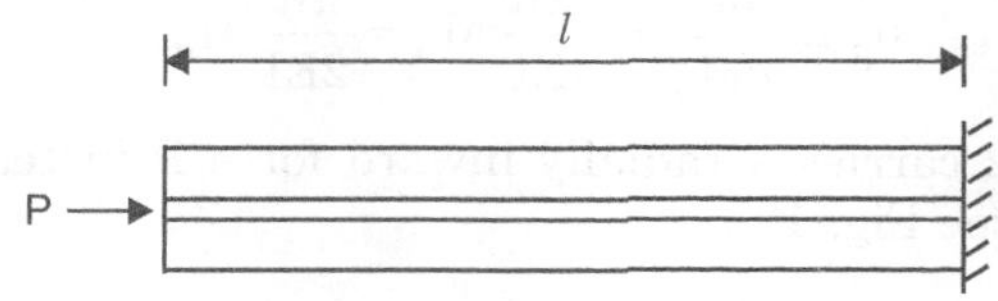

Fig. 11

Ans. Let F be the force in steel.

Force in concrete $= P - F$

Strain energy, $$U = \frac{F^2 l}{2A_sE_s} + \frac{(P-F)^2 l}{2A_cE_c}$$

$$\frac{\partial U}{\partial F} = 0$$

$$\frac{2Fl}{2A_sE_s} + \frac{2(P-F)(-1)l}{2A_cE_c} = 0$$

$$\frac{F}{100 \times 2 \times 10^5} + \frac{F}{900 \times 2 \times 10^4} - \frac{P}{900 \times 2 \times 10^4} = 0$$

$$F(0.1 + 0.11) = P(0.11)$$

$$F = \frac{0.11}{0.21}P$$

$$F = 0.523\ P.$$

21. A beam in the shape of a quadrant of a circle (Fig. 12) is acted on by a moment M_o at its free end. Find the rotation at the free end using Castigliano's theorem I.

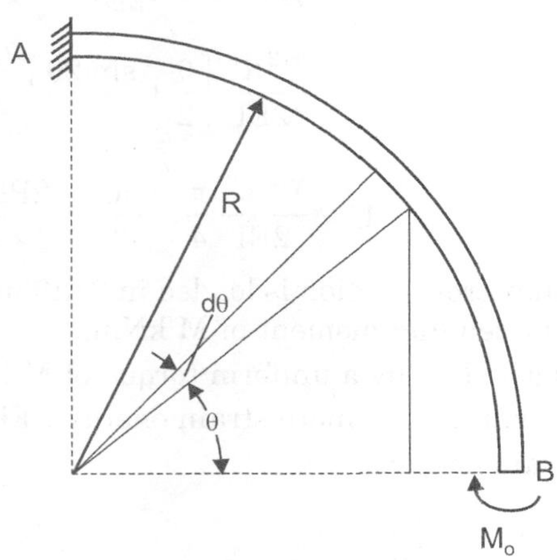

Fig. 12

Ans.

$$M = M_o$$

$$U = \int \frac{M_o^2\,ds}{2EI} = \frac{M_o^2 l}{2EI} = \frac{M_o^2}{2EI}\frac{\pi R}{2}$$

$$U = \frac{\pi R}{4EI} M_o^2$$

$$\theta_B = \frac{\partial U}{\partial M_o} = \frac{2\pi R}{4EI} M_o = \frac{\pi R}{2EI} M_o.$$

22. The beam in Fig. 12 carries a radially inward force P instead of M_o. Find the radial displacement at B (see Fig. 13).

Ans.

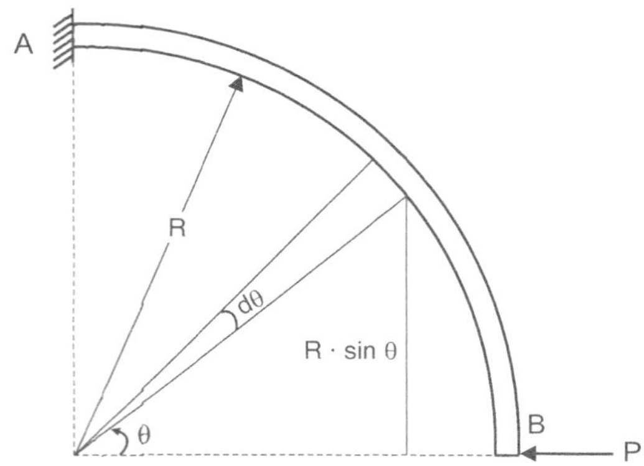

Fig. 13

The strain energy due to force P, $U_p = \int \frac{M^2 ds}{2EI}$

$$M_p = PR \sin\theta$$
$$ds = R\,d\theta$$

$$U_p = \int_0^{\pi/2} \frac{P^2R^2 \sin^2\theta\, R d\theta}{2EI}$$

$$U = \frac{P^2R^3}{2EI}\left[\frac{\theta}{2} + \frac{\sin 2\theta}{2}\right]_0^{\pi/2}$$

$$U = \frac{P^2R^3}{2EI}\frac{\pi}{4}; \quad \frac{\partial U}{\partial P} = \frac{2PR^3\pi}{2\times 4EI} = \frac{PR^3\pi}{4EI}.$$

23. A cantilever AB of circular cross-section is loaded in 2 different ways. In the first case it is subjected to a uniform bending moment of M kNm.

In the second case it is acted on by a uniform torque of M kNm.

In which of the 2 cases, will it store more strain energy ? Find the ratio of the energies.

Ans. For case 1, the strain energy

$$U_1 = \frac{M^2 l}{2EI}$$

For case 2, $$U_2 = \frac{M^2 l}{2GJ}$$

$$\frac{U_1}{U_2} = \frac{GJ}{EI} = \frac{E}{2(1+\nu)} \frac{\pi d^4}{32} \frac{1}{E} \frac{64}{\pi d^4}$$

$$\frac{U_1}{U_2} = \frac{1}{1+\nu}.$$

This is the ratio of the energies.

Hence U_2 (strain energy stored under torsion) is larger.

24. The strain energy in the beam (Fig. 14) due to M_o, $U_1 = \frac{M_o{}^2 l}{2EI}$. Strain energy due to P, $U_2 = \frac{P^2 l^3}{6EI}$. Prove that the strain energy due to both P and M_o acting at the free end is **not** equal to $U_1 + U_2$.

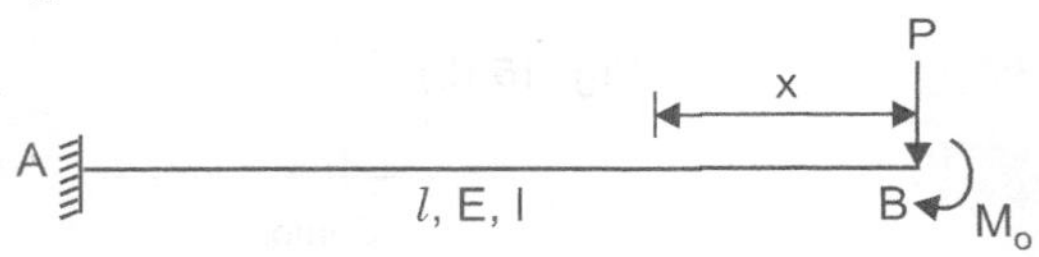

Fig. 14

Ans.

$$M_x = M_o + Px$$

$$U = \int_0^l \frac{(M_o + Px)^2\, dx}{2EI}$$

$$U = \int_0^l \frac{M_o{}^2\, dx}{2EI} + \int_0^l \frac{P^2 x^2\, dx}{2EI} + \int_0^l \frac{2M_o Px\, dx}{2EI}$$

$$= \frac{M_o{}^2 l}{2EI} + \frac{P^2 l^3}{6EI} + \frac{2M_o P l^2}{EI}$$

The first term is U_1

The second term is U_2.

The third term is not equal to zero.

Hence, $U > U_1 + U_2$.

25. Using the strain energy expression in question 24, find the rotation at B due to load P at B.

Ans.

$$U = \frac{M_o{}^2 l}{2EI} + \frac{P^2 l^3}{6EI} + \frac{M_o P l^2}{2EI}$$

$$\theta_B = \frac{\partial U}{\partial M_o} = \frac{2M_o l}{2EI} + 0 + \frac{Pl^2}{2EI}$$

Since only P acts, $M_o = 0$. $\therefore \; \theta_B = \frac{Pl^2}{2EI}$.

26. A simply supported member, hinged at both ends is acted upon by an axial force P as shown in Fig. 15(*a*). Find the reaction F at B using the principle of minimum strain energy.

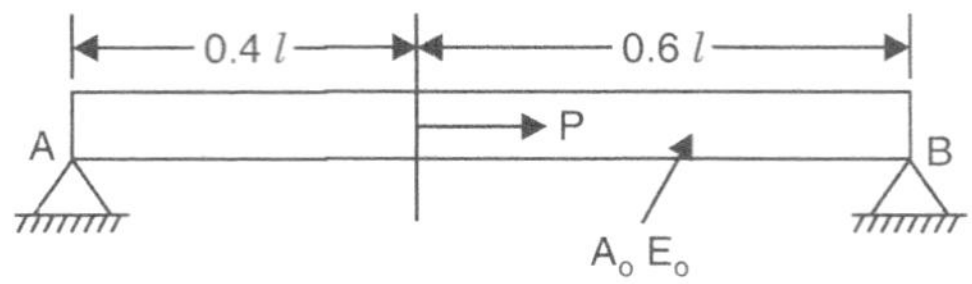

Fig. 15 (a)

Ans. Let us remove the support at B and introduce the support reaction F at P. Then the force diagram would be as in Fig. 15(*c*).

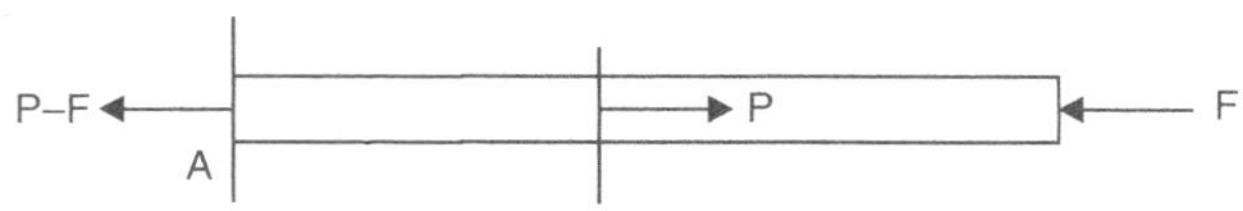

Fig. 15 (b)

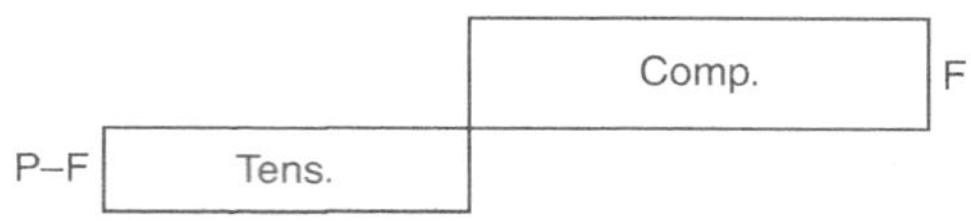

Fig. 15 (c)

The strain energy stored is given by

$$U = \frac{(P-F)^2 \times 0.4l}{2A_oE_o} + \frac{F^2 \times 0.6l}{2A_oE_o}$$

For U to be minimum, $$\frac{\partial U}{\partial F} = 0$$

$$\frac{2(P-F)(-1)\,0.4l}{2AE} + \frac{2F\,0.6l}{2AE} = 0$$

$$-0.4P + 0.4F + 0.6F = 0$$

$$F = 0.4P.$$

27. Using the strain energy expression in question 24, find the deflection at B due to M_o at B (see Fig. 14).

Ans. $$U = \frac{M_o^2 l}{2EI} + \frac{P^2l^3}{6EI} + \frac{M_oPl^2}{2EI}$$

$$\delta_B = \frac{\partial U}{\partial P} = 0 + \frac{2Pl^2}{6EI} + \frac{M_ol^2}{2EI}$$

Since P = 0, $\delta_B = \dfrac{M_ol^2}{2EI}$.

28. Compute the flexural strain energy stored in the cantilever frame in Fig. 16. Take EI = 1.

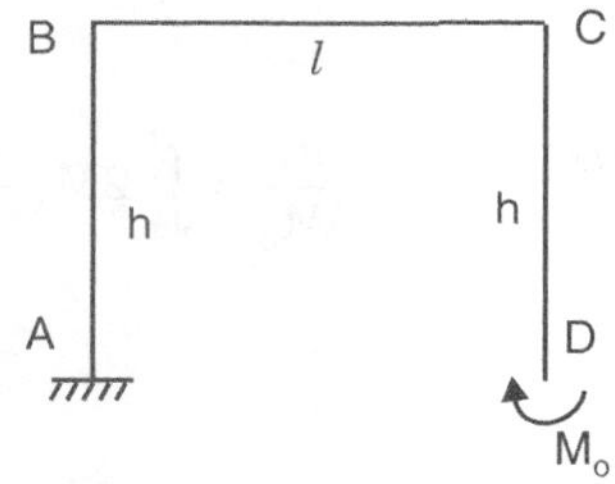

Fig. 16

Ans. Bending moment is M_o all over the frame.

Strain energy, $U = \dfrac{M_o^2 L}{2EI}$, when $L = 2h + l$

$$U = \frac{M_o^2(2h + l)}{2}.$$

29. In the frame in Fig. 16, what is the rotation at D ?

Ans.
$$U = \frac{M_o^2(2h + l)}{2}$$

$$\theta_D = \frac{\partial U}{\partial M_o} = M_o\,(2h + l),\ \text{since EI = 1.}$$

30. Find the moment at A of the propped cantilever in Fig. 17 by strain energy method.

Fig. 17

Ans. First introduce a hinge at A. Apply an external moment M_A at A and compute the strain energy U.

Since A is fixed $\dfrac{\partial U}{\partial M_o} = 0.$

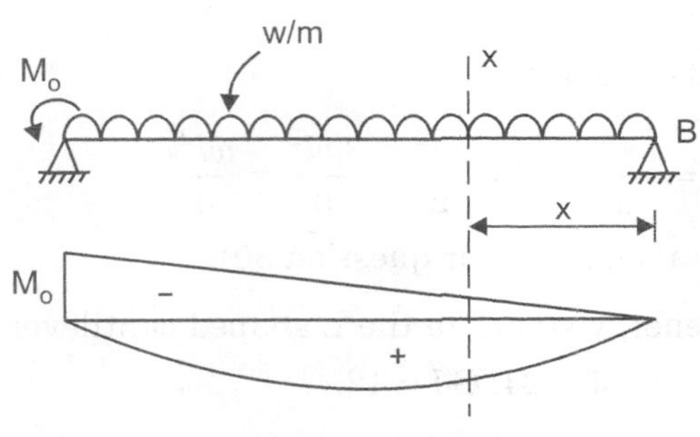

Fig. 18

$$M_x = \frac{wlx}{2} - \frac{wx^2}{2} - \frac{M_o x}{l}$$

$$U = \int \frac{M_x^2\, dx}{2EI}; \qquad \frac{\partial U}{\partial M_o} = \int_0^l 2M_x \frac{\frac{dM_x}{dM_o}dx}{2EI} = 0$$

$$\frac{dM_x}{dM_o} = \left(\frac{-x}{l}\right)$$

$$0 = \frac{1}{EI}\int_0^l \left(-\frac{wl}{2}\frac{x^2}{l} + \frac{wx^3}{2l} + \frac{M_o x^2}{l^2}\right) dx$$

$$= -\frac{wl^3}{6} + \frac{wl^3}{8} + \frac{M_o l}{3} = 0$$

$$M_o = \frac{wl^3}{24} \times \frac{3}{l} = \frac{wl^2}{8}.$$

31. For the propped cantilever in Fig. 17, find the prop reaction at B by strain energy method.

Ans. First remove the prop and introduce an upward force R at B.

$$M_x = Rx - \frac{wx^2}{2}$$

$$\frac{dU}{dR} = 0; \qquad \int_0^l M_x \frac{dM_x}{dR} dx = 0$$

$$\frac{dM_x}{dR} = x$$

$$\int_0^l \left(Rx^2 - \frac{wx^3}{2}\right) dx = 0$$

$$\frac{Rl^3}{3} - \frac{wl^4}{8} = 0; \quad R = \frac{3}{8}wl.$$

32. Find the moment at A of the propped cantilever in Fig. 17, from the fact that reaction at B is $\frac{3wl}{8}$.

Ans. Taking moments about A,

$$M_A = \frac{wl^2}{2} - Rl = \frac{wl^2}{2} - \frac{3wl^2}{8} = \frac{wl^2}{8}$$

(This can be used as a short-cut for question 30).

33. Compute the strain energy stored in the L shaped cantilever in Fig. 19.

EI = 24, GJ = 12.

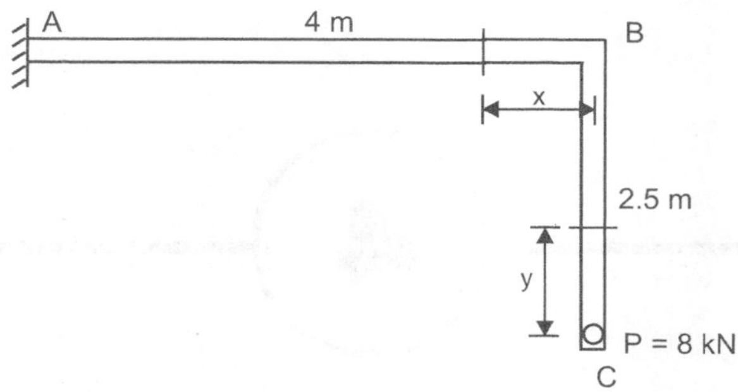

Fig. 19

Ans. Length BC will have only BM.

Length BA will have bending and torsion.

BC : $M_y = P_y = 8y$

$$U_1 = \int_0^{2.5} \frac{M_y^2\, dy}{2EI} = \int_0^{2.5} \frac{64y^2\, dy}{2 \times 24}$$

$$U_1 = \frac{64}{48} \frac{(2.5)^3}{3} = 6.94$$

BA : $M_x = P_x = 8x$

$$U_2 = \int_0^{4} \frac{M_x{}^2 dx}{2EI} = \int_0^{4} \frac{64x^2 dx}{48}$$

$$U_2 = \frac{64}{48} \frac{(4)^3}{3} = 28.44$$

$$T_x = P \times 2.5 = 20$$

$$U_3 = \frac{T^2 l}{2GJ} = \frac{400 \times 4}{2 \times 12} = 66.67$$

$$U = U_1 + U_2 + U_3 = 102.05 \text{ units.}$$

34. Find the twist at C due to an applied clockwise torque of 2 kNm at C in Fig. 19.

Ans. $T_o = 2$

Torsion on BC = T_o

Strain energy due to T_o, $U_1 = \dfrac{T_o{}^2 l}{2GJ} = \dfrac{T_o{}^2 \times 2.5}{2 \times 12}$

T_o will produce uniform bending on AB.

Strain energy due to T_o, $U_2 = \dfrac{M^2 l}{2EI}$

$$U = U_1 + U_2 = \frac{T_o{}^2 \times 2.5}{24} + \frac{T_o{}^2 \times 4}{48} = 0.1875\, T_o{}^2$$

$$\theta_C = \frac{\partial U}{\partial T_o} = 0.1875 \times 2T_o = 0.375\, T_o$$

$$\theta_C = 0.375 \times 2 = 0.75 \text{ units.}$$

4

Slope-Deflection Method

1. What are the assumptions made in slope-deflection method ?

   Ans. Assumptions made in slope-deflection method are

   (*i*) between each pair of supports the beam section is constant.

   (*ii*) the joint in a structure may rotate or deflect as a whole, but the angles between the members meeting at that joint remain the same.

2. How many slope-deflection equations are available for a two span continuous beam ?

   Ans. There will be 4 Nos. of slope-deflection equations, two for each span.

3. What is the moment at a hinged end of a simple beam ?

   Ans. Moment at the hinged ends of a simple beam = zero.

4. What are the quantities in terms of which the unknown moments are expressed in slope-deflection method ?

   Ans. In slope-deflection method, unknown moments are expressed in terms of

   (*i*) slopes (θ) and (*ii*) deflections (Δ)

5. The beam shown in Fig. 1 is to be analysed by slope-deflection method. What are the unknowns and, to determine them, what are the conditions used?

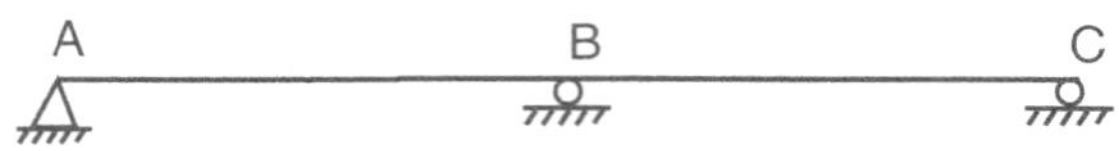

Fig. 1

   Ans. Unknowns : θ_A, θ_B, θ_C

   Equilibrium equations used : $M_{AB} = 0$

   $$M_{BA} + M_{BC} = 0$$

   $$M_{CB} = 0.$$

6. How do you account for sway in slope-deflection method for portal frames ?

   Ans. Because of sway, there will be rotations in the vertical members of a frame. This causes moments in the vertical members. To account for this, besides the equilibrium, one more equation namely shear equation connecting the joint-moments is used.

7. Write down the equation for sway correction for the portal frame shown in Fig. 2.

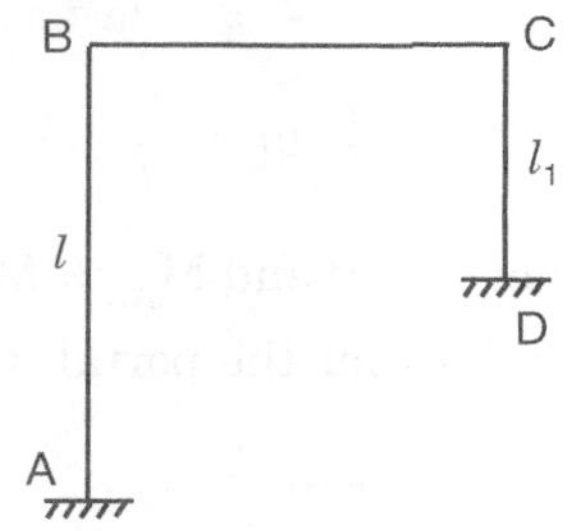

Fig. 2

Ans. The shear equation (sway correction equation) is

$$\frac{M_{AB} + M_{BA}}{l} + \frac{M_{CD} + M_{DC}}{l_1} = 0.$$

8. For the portal frame shown in Fig. 3, write down the horizontal shear equation.

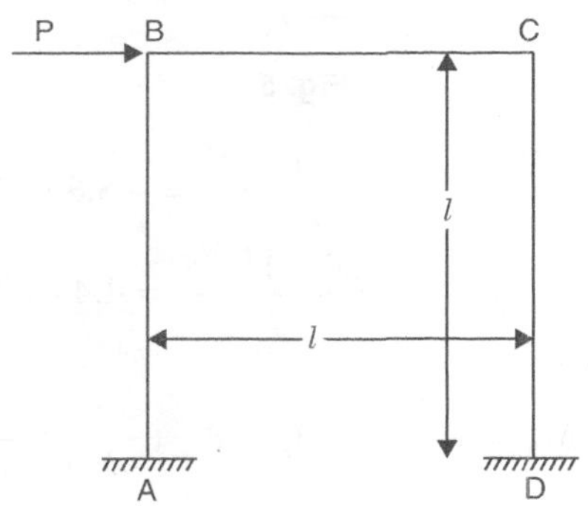

Fig. 3

Ans. The shear equation is

$$\frac{M_{AB} + M_{BA}}{l} + \frac{M_{CD} + M_{DC}}{l} + P = 0.$$

9. In Fig. 3, if the support D were hinged, what will be the shear equation ?

Ans. $\frac{M_{AB} + M_{BA}}{l} + \frac{M_{CD}}{l} + P = 0.$

10. Write down the slope-deflection equation for a fixed end support.

Ans. Assuming a beam as below :

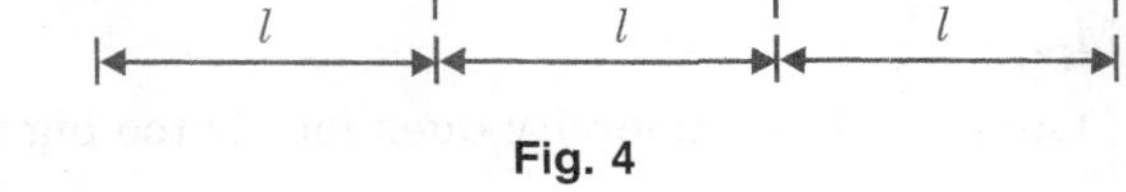

Fig. 4

The slope-deflection equation for end A is

$$M_{AB} = M_{FAB} + \frac{2EI}{l}\left(2\theta_A + \theta_B + \frac{3\Delta}{l}\right)$$

$$= M_{FAB} + \frac{2EI}{l}\left(\theta_B + \frac{3\Delta}{l}\right) \qquad \text{(Since } \theta_A = 0\text{)}$$

If there is no support settlement, $\Delta = 0$ and $M_{AB} = M_{FAB} + \frac{2EI}{l}(\theta_B)$

11. Write the slope-deflection equations for the portal frame in Fig. 5.

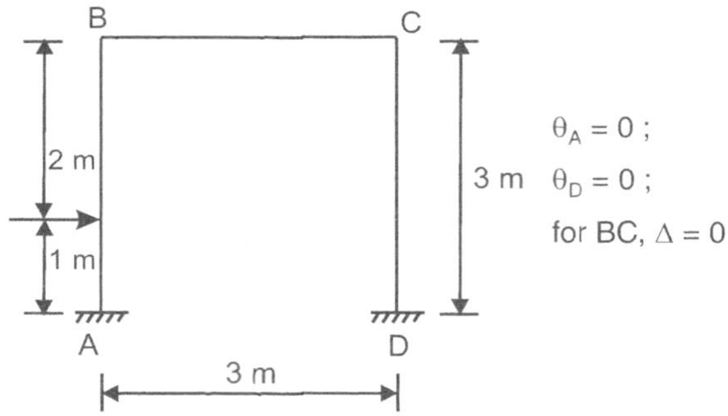

Fig. 5

Ans.

$$M_{FAB} = -\frac{Wab^2}{l^2} = \frac{20 \times 1 \times 2^2}{3^2} = -8.89 \text{ kNm}$$

$$M_{FBA} = +\frac{Wa^2b}{l^2} = +\frac{20 \times 1^2 \times 2}{3^2} = 4.44 \text{ kNm}$$

Slope-deflection equations :

$$M_{AB} = M_{FAB} + \frac{2EI}{l}\left(2\theta_A + \theta_B + \frac{3\Delta}{l}\right) = -8.89 + \frac{2EI}{l}\left(\theta_B - \frac{3\Delta}{3}\right)$$

$$M_{AB} = -8.89 + \frac{2EI}{l}(\theta_B - \Delta) \qquad \text{...(1)}$$

$$M_{BA} = M_{FBA} + \frac{2EI}{l}\left(2\theta_B + \theta_A + \frac{3\Delta}{l}\right)$$

$$M_{BA} = 4.44 + \frac{2EI}{3}(2\theta_B - \Delta) \qquad \text{...(2)}$$

$$M_{BC} = \frac{2EI}{3}(2\theta_B + \theta_C) \qquad \text{...(3)}$$

$$M_{CB} = \frac{2EI}{3}(2\theta_C + \theta_B) \qquad \text{...(4)}$$

$$M_{CD} = 0 + \frac{2EI}{3}\left(2\theta_C + \theta_D + \frac{3\Delta}{l}\right)$$

$$M_{CD} = \frac{2EI}{3}(2\theta_C - \Delta) \qquad \text{...(5)}$$

$$M_{DC} = \frac{2EI}{3}(\theta_C - \Delta) \qquad \text{...(6)}$$

Authors comment : Question 11 (University question) is too big for 'short questions'.

12. What are shear equations ?

Ans. When a building frame is subjected to side sway, the sum of the internal shear forces in all the columns at any level, expressed in terms of the column moments, must be equal to the total external lateral force applied to the frame above that level, to satisfy the condition $\Sigma H = 0$. This condition gives rise to a set of equations known as shear equations.

13. How many shear equations can be written for an '*n*' storeyed building ? What is its use ?

Ans. One shear equation can be written for each storey. Hence the number of shear equations is equal to the number of storeys in a building.

Use : Gives additional equation/s to solve for the additional unknown/s resulting from the sway of a portal frame.

14. Write the equilibrium equations required for the frame shown in Fig. 6.

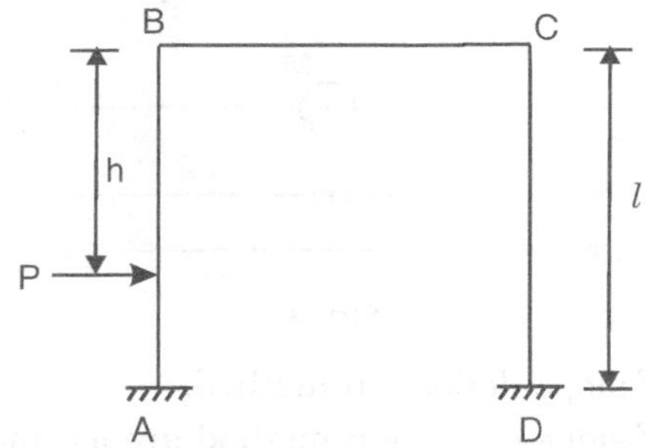

Fig. 6

Ans. Unknowns : θ_B, θ_C, D

Equilibrium equations : At B, $M_{BA} + M_{BC} = 0$

At C, $M_{CB} + M_{CD} = 0$

Shear equation : $\dfrac{M_{AB} + M_{BA} - Ph}{l} + \dfrac{M_{CD} + M_{DC}}{l} + P = 0$

15. Who introduced slope-deflection method of analysis ? *(Anna University, Nov. 2003)*

Ans. Slope-deflection method was introduced by Prof. George A. Maney in 1915.

16. Write down the general slope-deflection equations and state what each term represents. *(Anna University, Nov. 2003)*

Ans.

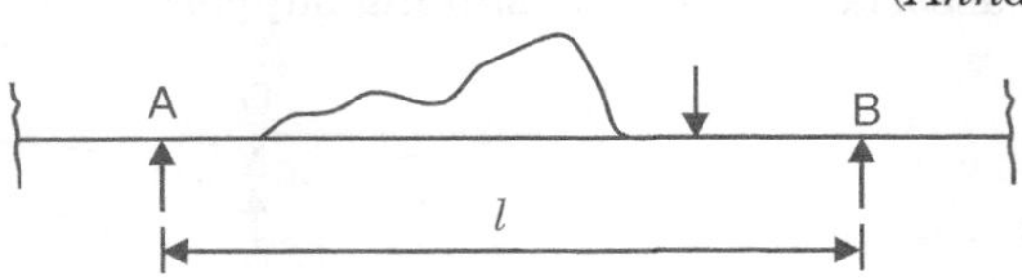

Fig. 7

General slope-deflection equations :

$$M_{AB} = M_{FAB} + \frac{2EI}{l}\left(2\theta_A + \theta_B + \frac{3\Delta}{l}\right)$$

$$M_{BA} = M_{FBA} + \frac{2EI}{l}\left(2\theta_B + \theta_A + \frac{3\Delta}{l}\right)$$

M_{FAB}, M_{FBA} — Fixed end moment at A and B respectively due to the given loading

θ_A, θ_B — Slopes at A and B respectively

Δ — Sinking of support A with respect to B.

17. Mention any three reasons due to which sway may occur in portal frames. *(Anna University, Nov. 2003)*

Ans. Sway in portal frames may occur due to

(*i*) Unsymmetry in geometry of the frame

(*ii*) Unsymmetry in loading or

(*iii*) Settlement of one end of a frame.

18. How many slope-deflection equations are available for each span ?

Ans. Two numbers of slope-deflection equations are available for each span, defining the moment at each end of the span.

19. Write the fixed end moments for a beam carrying a central clockwise moment.

Ans. Fixed end moments : $M_{FAB} = M_{FBA} = \frac{M}{4}$

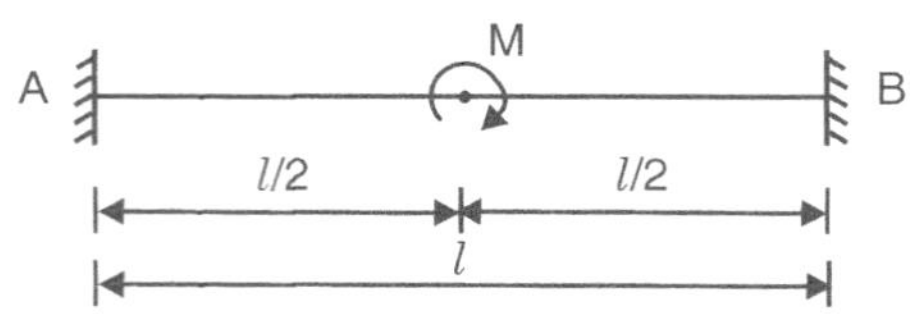

Fig. 8

20. State the limitations of slope-deflection method.

Ans. The limitations of slope-deflection method are as under

(*i*) It is not easy to account for varying member sections

(*ii*) It becomes very cumbersome when the unknown displacements are large in number.

21. Why is slope-deflection method called a 'displacement method'?

Ans. In slope-deflection method, displacements (like end slopes and settlements) are treated as unknowns and hence the method is a 'displacement method'.

22. Define 'degrees of freedom'.

Ans. In a structure, the number of independent joint displacements that the structure can undergo are known as degrees of freedom.

23. In a continuous beam, one of the supports sinks. What will happen to the span and support moments associated with the sinking support ?

Ans.

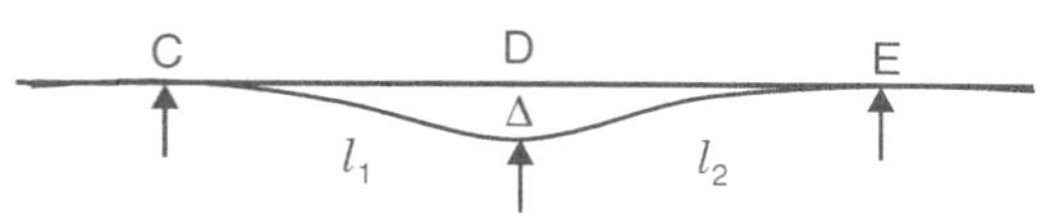

Fig. 9

Referring to Fig. 9, let support D sink by Δ. This will not affect span moments. Fixed end moments (support moments) will get developed as under

$$M_{FCD} = M_{FDC} = -\,6EI\,\frac{\Delta}{l_1^2}$$

$$M_{FDE} = M_{FED} = 6EI\,\frac{\Delta}{l_2^2}$$

24. A rigid frame is having totally 10 joints including support joints. Out of slope-deflection and moment distribution methods, which method would you prefer for analysis ? Why ?

Ans. Obviously moment distribution method is preferable.

If we use slope-deflection method, there would be ten (or more) unknown displacements and an equal number of equilibrium equations. In addition there would be 2 unknown support moments per span and the same number of slope-deflection equations. Solving them can be a formidable task.

25. What is the basis on which a sway equation is formed for a structure ?

Ans. Sway is dealt with in slope-deflection method (like in moment distribution method) by considering the horizontal equilibrium of the whole frame taking into account the shears at the base level of columns and external horizontal forces.

Thus in Fig. 6, the 'shear condition' reads as,

$$\frac{M_{AB} + M_{BA} - Ph}{l} + \frac{M_{DC} + M_{DC}}{l} + P = 0$$

26. Write the equilibrium equation for support B of the beam shown in Fig. 10. Express the values involved in terms of θ_B the rotation at B.

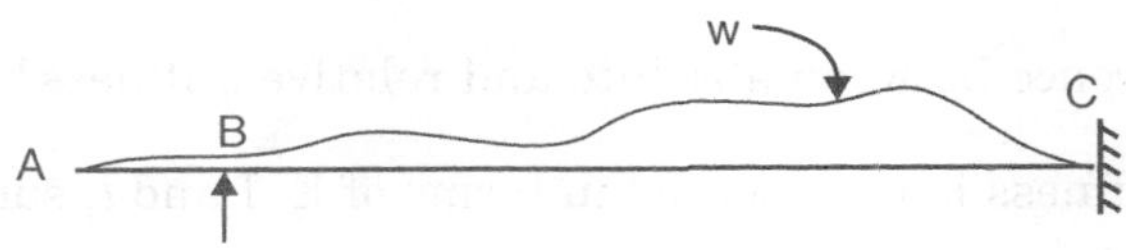

Fig. 10

Ans. (*i*) Equilibrium equation

$$M_{BA} + M_{BC} = 0$$

(*ii*) M_{BC} is given by the slope-deflection equation

$$M_{BC} = M_{FBC} + \frac{2EI}{l} 2\theta_B$$

(*iii*) M_{BA} is the moment due to loading on the overhang BA.

27. List the unknown displacements involved in solving the frame in Fig. 11.

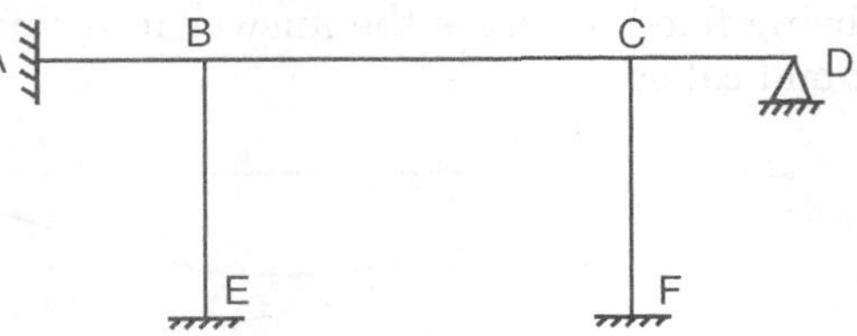

Fig. 11

Ans. The unknown displacements are, θ_B, θ_C and θ_D.

28. What are the unknown displacements involved in solving the frame in Fig. 12.

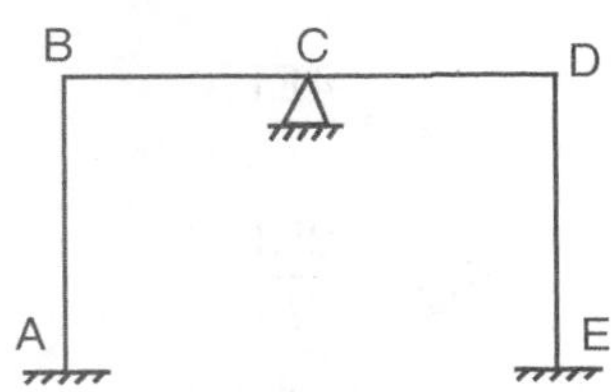

Fig. 12

Ans. The unknown displacements are θ_B, θ_D and θ_C.

5

Moment Distribution Method and Naylor's Method

1. What is the difference between absolute and relative stiffness?

Ans. Absolute stiffness is represented in terms of E, I and l, such as $\frac{4EI}{l}$.

Relative stiffness is represented in terms of I and l, omitting the constant E. Relative stiffness is the ratio of stiffnesses of 2 or more members at a joint.

2. In a member AB, if a moment of –10 kNm is applied at A, what is the moment carried over to B ?

Ans. Carry over moment = Half the applied moment

$\therefore$ Carried over moment to B $= -\frac{10}{2} = -5$ kNm

3. What are the moments induced in a beam member, when one end is given a unit rotation, the other end being fixed. Express the answer in terms of l, E and I. What is the moment at the near end called ?

Ans.

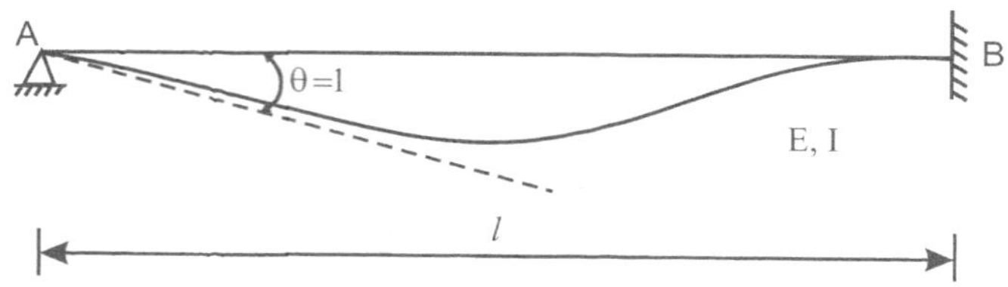

Fig. 1

When $\theta_A = 1$, (Fig. 1)

$$M_{AB} = \frac{4EI}{l}$$

$$M_{BA} = \frac{2EI}{l}$$

M_{AB} is the stiffness of AB at B.

4. Define the term "Distribution factor".

Ans. When a moment M is applied to a rigid joint 'O', where two or more members meet, all the members at the joint will undergo the same rotation θ. Different members (meeting at the joint) may require different values of moment to have the same rotation. This depends upon the stiffness of the members. The factor by which M is to be multiplied to get the end moment shared by any member is defined as the distribution factor.

5. A beam ABC is fixed at A and simply supported at B and C. AB = BC = l. Flexural rigidities of AB and BC are 2EI and EI, respectively. Find the distribution factors at joint B if no moment is to be transferred to support C.

Ans.

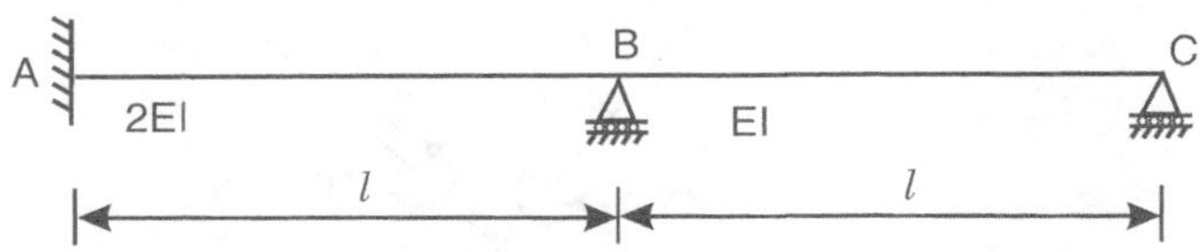

Fig. 2

Joint B :

Relative stiffness : $\frac{I_1}{l} = \frac{2I}{l}$ for BA $\quad k_{BA} = 2$

$\frac{3}{4} \times \frac{I_2}{l} = \frac{3I}{4l}$ for BC $\quad k_{BC} = \frac{3}{4} = 0.75$

Distribution factors :

DF for BA : $\frac{k_{BA}}{k_{BA} + k_{BC}} = \frac{2}{2 + 0.75} = \frac{8}{11} = 0.727$

DF for BC : $\frac{k_{BC}}{k_{BC} + k_{BA}} = \frac{0.75}{2 + 0.75} = \frac{3}{11} = 0.273$

6. What is the sum of distribution factors at a joint ?

Ans. Sum of distribution factors at a joint = 1.

7.

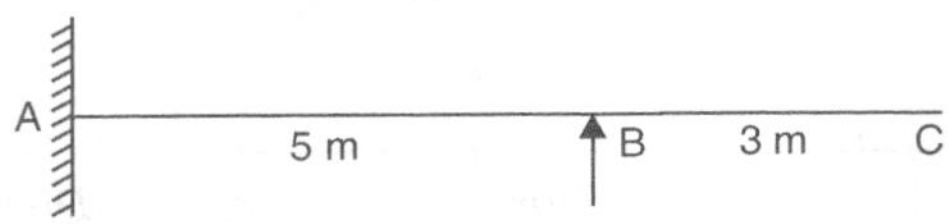

Fig. 3

For the joint B of the beam shown in Fig. 3, determine the distribution factors.

Ans. No moment will go to BC. Hence DF are 1 and 0 for BA and BC.

8. Explain "carry over factor" with a sketch.

Ans.

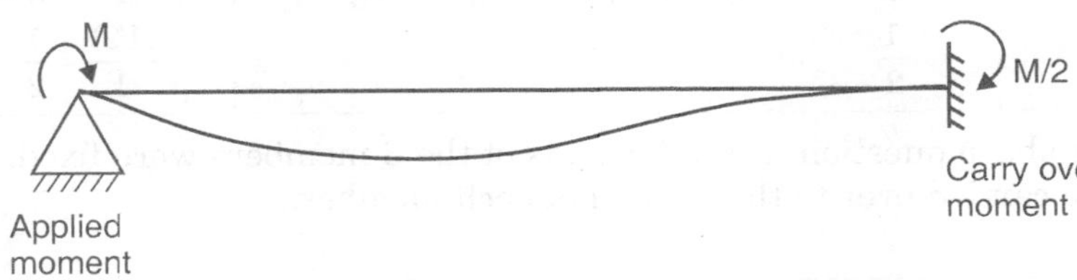

Fig. 4

Carry over factor is the ratio of moment induced at the far end and the moment applied at the near end.

Considering prismatic beams as shown in Fig. 4

Carry over factor, C.O.F. = $\frac{M/2}{M} = \frac{1}{2}$

If the far end is simply supported, there is no carry over moment. Hence the carry over factor is zero.

9. At a certain stage of moment distribution analysis, three members meeting at a rigid joint carry moments of 5 kNm (clockwise), 10 kNm (clockwise) and 9 kNm (counter clockwise). What moment should be applied at the joint O for achieving equilibrium ?

Ans.

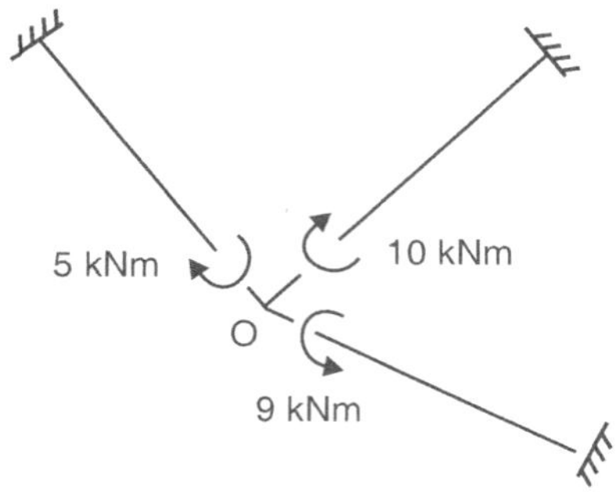

Fig. 5

Unbalanced moment = 5 (clockwise) + 10 (clockwise) − 9 (counter clockwise)

= 6 kNm (clockwise)

Moment to be applied at O for equilibrium

= 6 kNm (counter clockwise)

10. Distribute a moment of 6 kNm at a joint among three members whose relative stiffnesses are $\frac{1}{6}, \frac{1}{3}$ and $\frac{1}{2}$.

Ans.

Member	*Relative stiffness*	*Sum*	*Distribution factor*	*Moment in member (kNm)*
1	$\frac{1}{6}$		$\frac{(1/6)}{1} = \frac{1}{6}$	$\frac{1}{6} \times 6 = 1$
2	$\frac{1}{3}$	$\frac{1}{6} + \frac{1}{3} + \frac{1}{2} = \frac{1+2+3}{6} = \frac{6}{6} = 1$	$\frac{(1/3)}{1} = \frac{1}{3}$	$\frac{1}{3} \times 6 = 2$
3	$\frac{1}{2}$		$\frac{(1/2)}{1} = \frac{1}{2}$	$\frac{1}{2} \times 6 = 3$

11. In the above question, if the far ends of the 3 members were fixed, how much moment will be carried over to the far end of each member.

Ans.

Member No.	*Moment shared*	*Moment carried over to the other end*
1	1 kNm	0.5 kNm
2	2 kNm	1.0 kNm
3	3 kNm	1.5 kNm

12. Define the term "sway".

Ans. Sway is the lateral movement of joints in a portal frame due to the unsymmetry in dimensions, loads, moments of inertia, end conditions, etc.

13. What are the situations wherein sway will occur in portal frames ?

Ans. Situations in which sway will occur in portal frames :

(*i*) Eccentric or unsymmetric loading

(*ii*) Unsymmetrical geometry

(*iii*) Different end conditions of the columns

(*iv*) Non-uniform section of the members

(*v*) Unsymmetrical settlement of supports

(*vi*) A combination of the above.

14. What is the ratio of sway moments at column heads when one end is fixed and the other end hinged ? Assume that the length and M.I. of both legs are equal.

Ans. Assuming the frame to sway to the right by *d* (see Fig. 6).

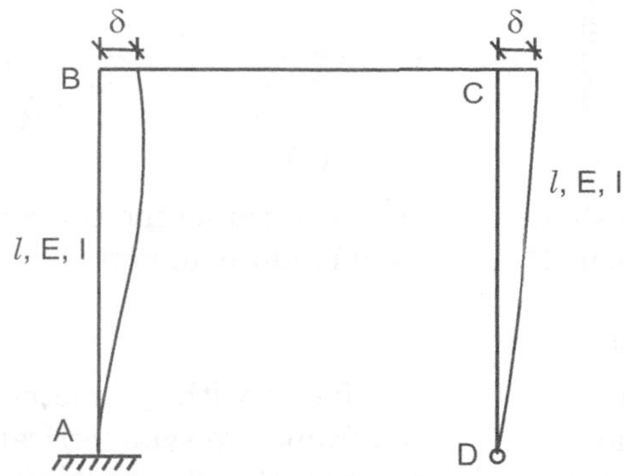

Fig. 6

Ratio of sway moments, $$\frac{M_{BA}}{M_{CD}} = \frac{-\left(\frac{6EI\delta}{l^2}\right)}{-\left(\frac{3EI\delta}{l^2}\right)} = 2$$

15. How will you account for the sway of a portal frame with one inclined leg ?

Ans. Refer to Fig. 7.

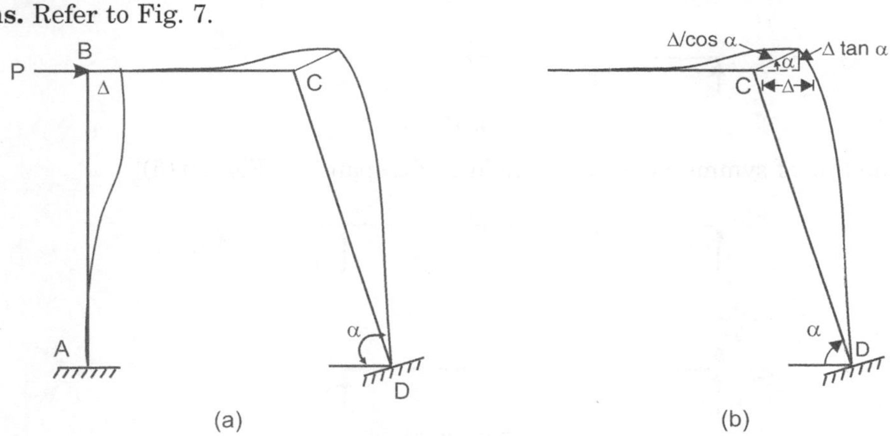

Fig. 7

In the case of portal frames with inclined members, the translation and rotation of the beam BC will also be taken into account in the calculation of correction moments. Thus if AB has a sway of Δ, DC will sway by $\Delta/\cos\alpha$ and BC will sway by $\Delta \tan\alpha$.

16. A beam is fixed at its left end and simply supported at right. The right end sinks to a lower level by a distance 'Δ' with respect to the left end. Find the magnitude and direction of the reaction at right end if l is the beam length and EI, the flexural rigidity.

Ans.

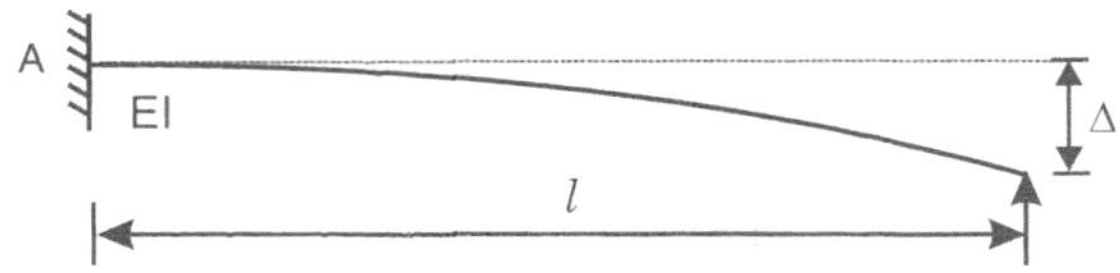

Fig. 8

$$M_A \text{ (due to sinking of B)} = \frac{3EI\Delta}{l^2}$$

17.

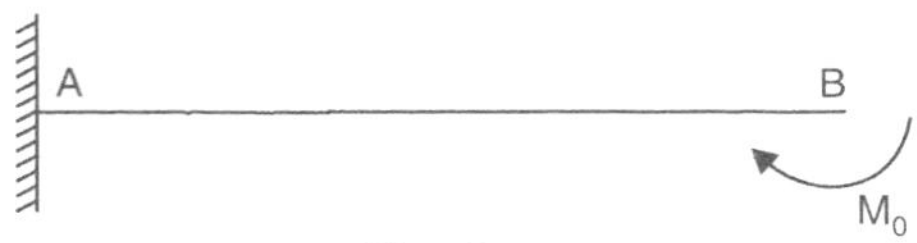

Fig. 9

For the cantilever AB shown in Fig. 9, determine the carry over factor from B to A.

Ans. A clockwise moment M_0 at B would induce an anti-clockwise moment M_0 at support A.

Hence the carry over factor is – 1.

18. How do you simplify the analysis of a frame with symmetry of geometry and loading ?

Ans. If the geometry and loading of a frame are symmetrical, we take advantage of the same and only the half of the frame is analysed. Here 2 possible cases will arise.

(*i*) There is a support at the line of symmetry [Fig. 10 (*a*)]. This is simplified by treating the line of symmetry as a fixed end.

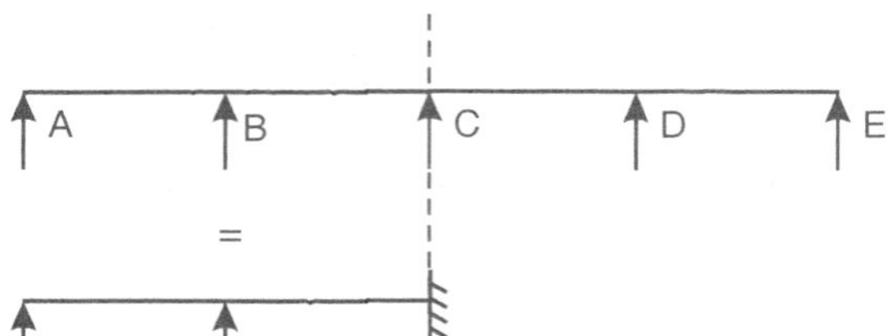

Fig. 10 (a)

(*ii*) The line of symmetry is at the middle of a span [see Fig. 10 (*b*)]

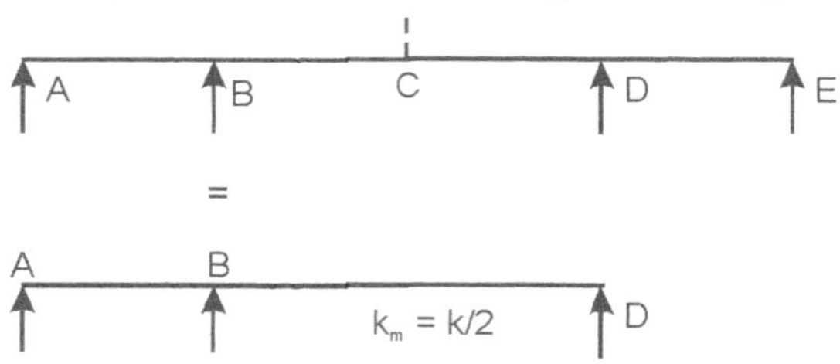

Fig. 10 (b)

This is simplified by taking only half the value of the stiffness of BD and not carrying over any moment to or from D.

19. Taking advantage of symmetry, find the final moment M_{BC} in the beam shown in Fig. 11.

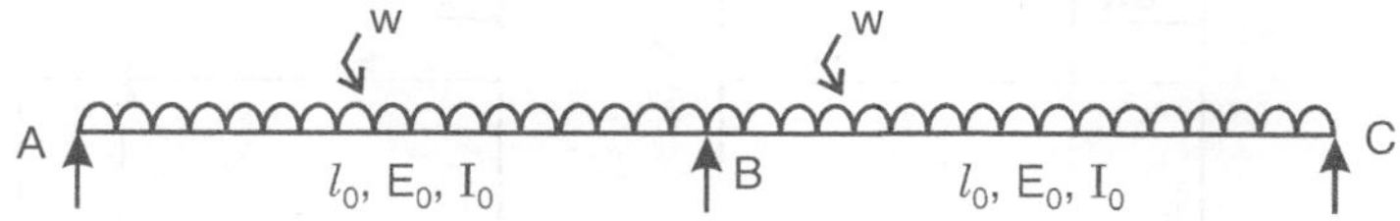

Fig. 11

Ans.

Fig. 12

The beam is equivalent to the one shown in Fig. 12.

$$-M_{FAB} = M_{FBA} = \frac{wl_0^2}{12}\text{; To balance } -\frac{wl_0^2}{12} \text{ at A, we apply } \frac{wl_0^2}{12}.$$

Carry over from A to B = $\dfrac{wl_0^2}{24}$

$$M_{BA} = \frac{wl_0^2}{12} + \frac{wl_0^2}{24} = \frac{3wl_0^2}{24} = \frac{wl_0^2}{8}.$$

20. Analyse the continuous beam shown in Fig. 13 by taking advantage of symmetry.

	A	B		C		D
FEM	– 18	18	– 24	24	– 16	+ 16
$\frac{I}{l}$ (Relative stiffness)		1.6	1.2	1.2	1.6	

Fig. 13. FEMs, *I/l* values

Ans. The simplified beam will have only half the stiffness for BC. BA will have $\dfrac{3}{4}\dfrac{I}{l}$ as modified stiffness. Thus we have a beam as shown in Fig. 14.

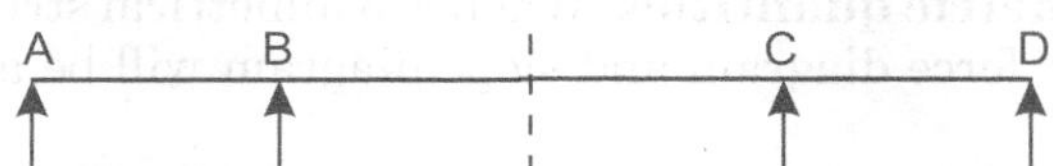

	A	B		C		D
Modified stiffness		1.2	0.6			
FEM	– 18	18	– 24			
Balance, CO	+ 18	9				
Balance		–2	–1			
Final moments	0	25	– 25			
Final moments (full beam)	0	25	– 25	25	– 25	0

Fig. 14

21. Sketch a structure symmetrical about the vertical axis with symmetric and anti-symmetric loadings.

Ans.

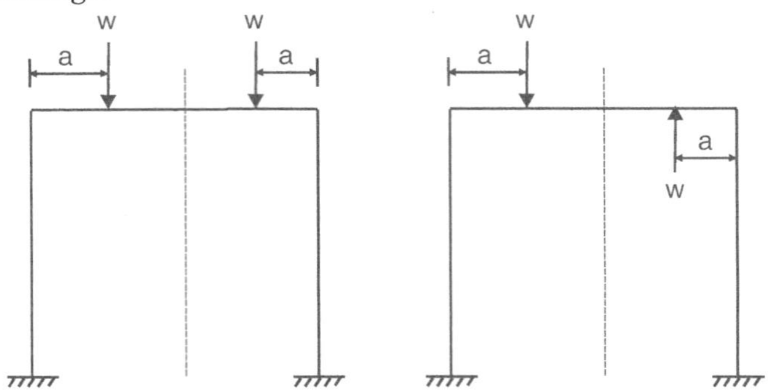

(a) Symmetric loading (b) Anti-symmetric loading

Fig. 15. Structure Symmetrical about vertical axis.

22. What are symmetric and anti-symmetric quantities in a symmetrically loaded beam?

Ans. Symmetric quantities. When a symmetrical structure is loaded with symmetrical loading, the bending moment and deflected shape will be symmetrical about the centre. Bending moment and deflection are symmetrical quantities (vide Fig. 16).

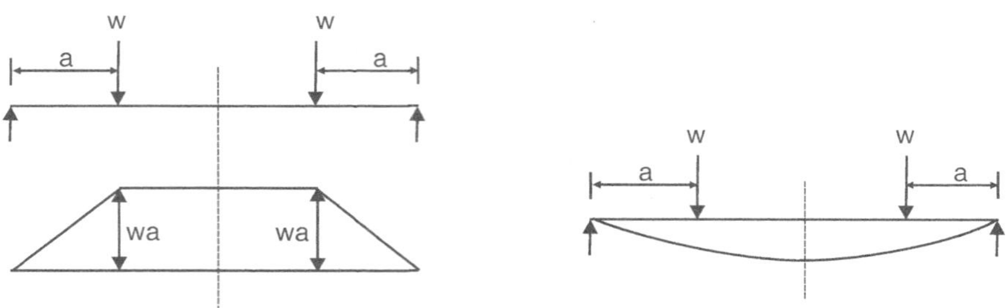

(a) Bending moment diagram (b) Deflection diagram

Fig. 16

Anti-symmetric quantities. When a symmetrical structure is subjected to a symmetric loading, the shear force diagram and slope diagram will be anti-symmetric (see Fig. 17).

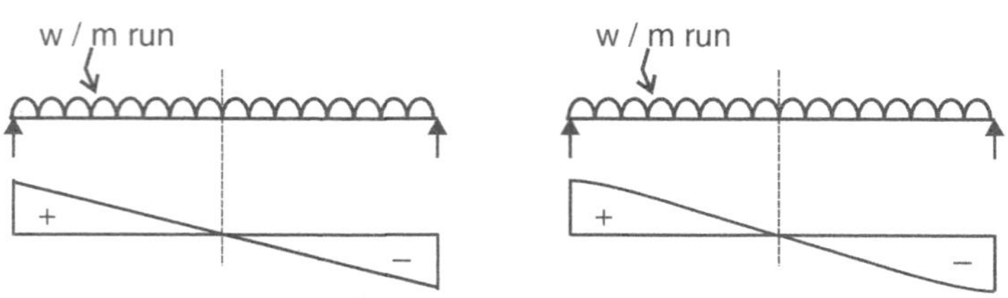

(a) Shear force diagram (b) Slope diagram

Fig. 17

Shear force and slope are anti-symmetric quantities.

23. For the structure loaded as shown in Fig. 18 (*a*), what is the deflection at E.

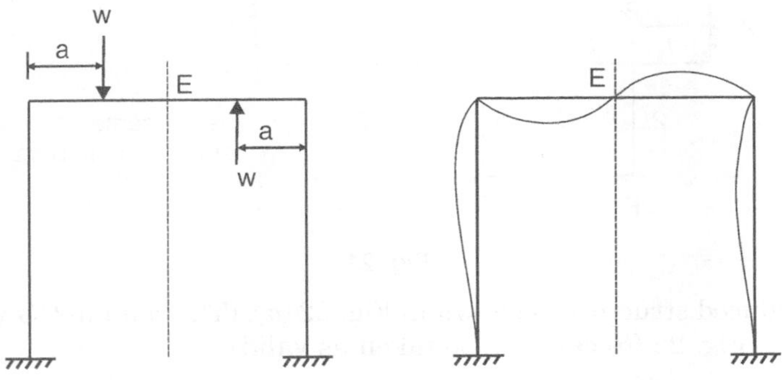

(a) Given structure and loading (b) Deflected shape

Fig. 18

Ans. Deflection at E is zero [vide Fig. 18 (*b*)]. (For anti-symmetric loading, symmetric quantities will be zero at the point of symmetry).

24. Express the following loading case as combination of symmetrical and anti-symmetrical loadings (vide Fig. 19)

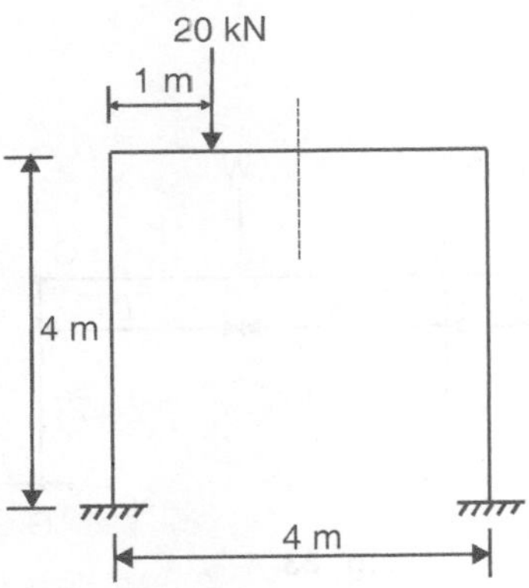

Fig. 19

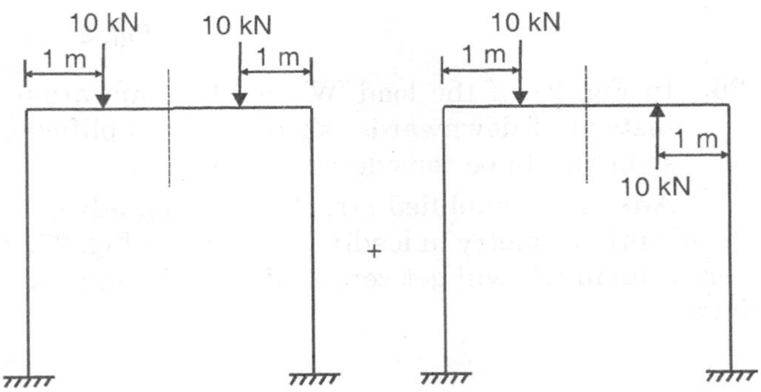

(a) Symmetric loading (b) Anti-symmetric loading

Fig. 20

Ans. Fig. 20 shows the symmetric and anti-symmetric parts. When these are added, we get the given loading.

25. For the structure in Fig. 21, take advantage of symmetry in the structure and the loading and show the reduced structure for analysis. Clearly indicate the support condition at the point of symmetry.

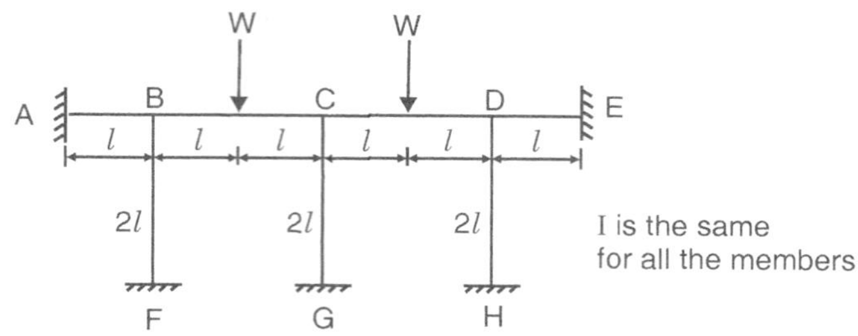

Fig. 21

Ans. The reduced structure is shown in Fig. 22 (*a*). (The column CG will get no bending moment. Therefore, Fig. 22 (*b*) can also be taken as valid).

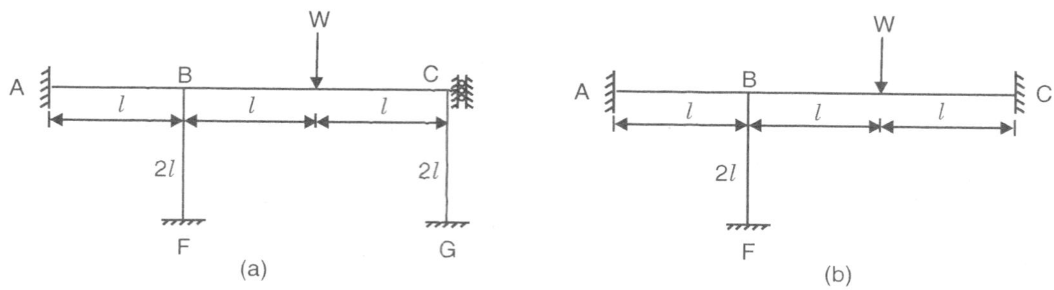

Fig. 22

26. In Fig. 21, if the load 'W' on CD is upwards instead of downwards, sketch the simplified structure to be considered for analysis.

Ans. The simplified structure taking advantage of anti-symmetry in loading is shown in Fig. 23. (Here column CG will get zero load). C will act as a hinge.

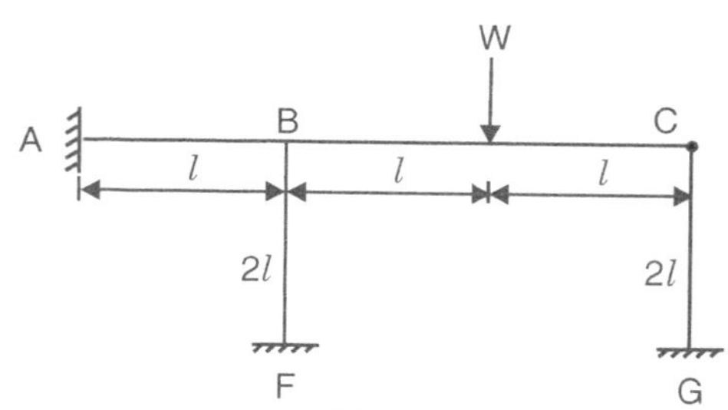

Fig. 23

27. Express the general loading case in Fig. 24 as combinations symmetrical and antisymmetrical loading cases.

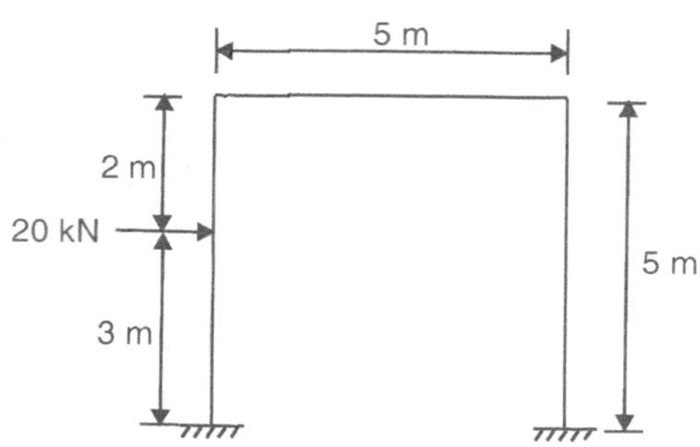

Fig. 24

Ans. Fig. 25 shows the symmetric and anti-symmetric components of the given loading.

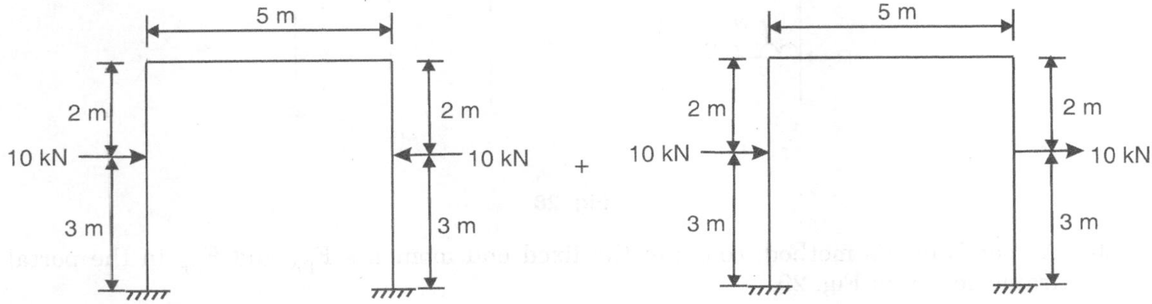

(a) Symmetric loading (b) Anti-symmetric loading

Fig. 25

28. Express the following loading case as combination of symmetrical and anti-symmetrical loading cases (vide Fig. 26).

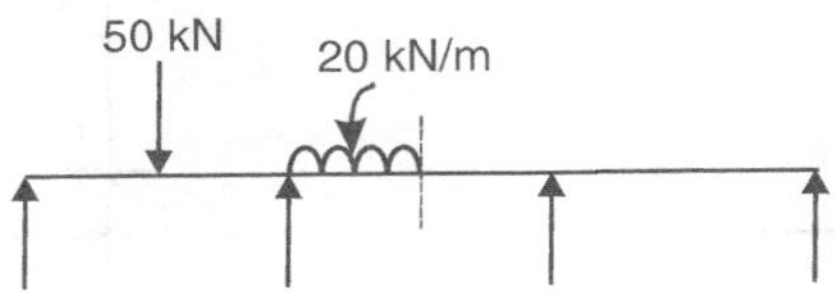

Fig. 26

Ans. Fig. 27 shows the symmetric and anti-symmetric components of the give loading.

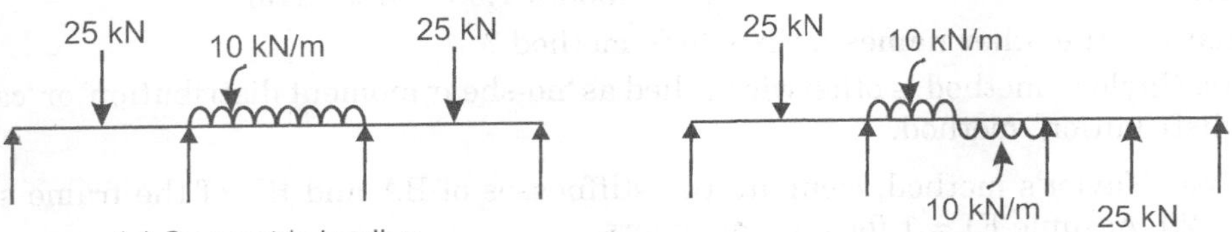

(a) Symmetric loading (b) Anti-symmetric loading

Fig. 27

29. In the following figures [Fig. 28 (*a*) to (*f*)], pick out the cases having sway.

Ans. Cases (*b*) and (*d*) will have sway. Cases *a*, *c*, *e* and *f* will not have sway.

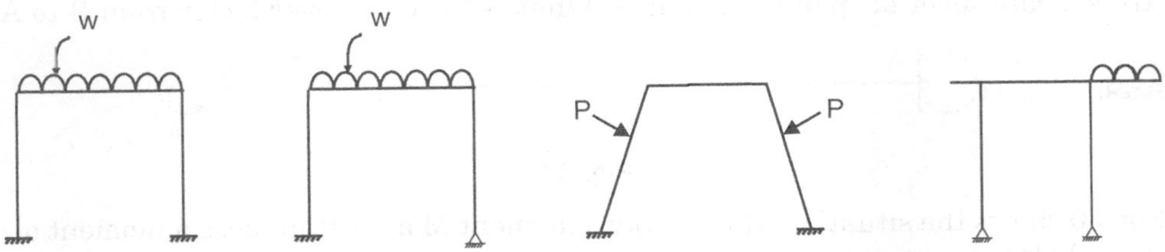

(a) (b) (c) (d)

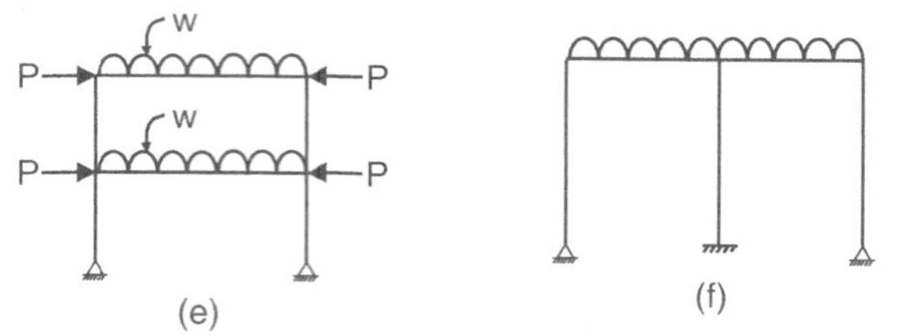

Fig. 28

30. As per Naylor's method, compute the fixed end moments F_{BA} and F_{BE} in the portal frame shown in Fig. 29.

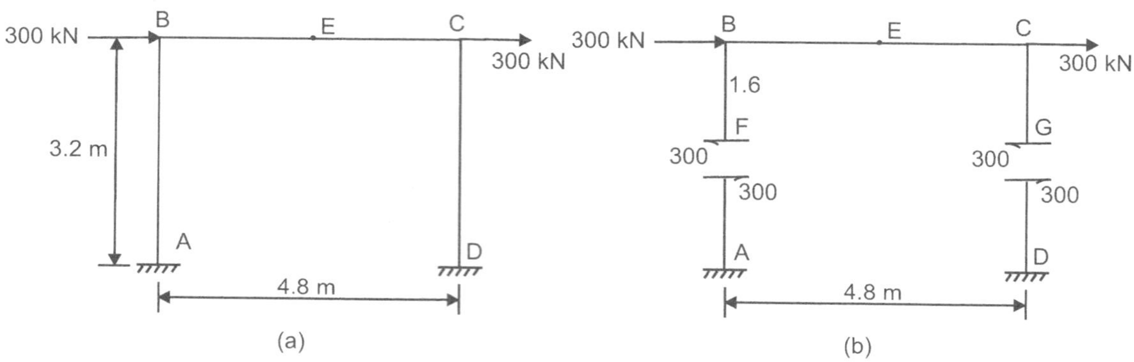

Fig. 29

Ans.

$$F_{BA} = -\,300 \times 1.6 = -\,480 \text{ kNm}$$

31. What are the other names for Naylor's method ?

Ans. Naylor's method is otherwise called as 'no-shear moment distribution' or 'cantilever moment distribution' method.

32. As per Naylor's method, compute the stiffnesses of BA and BE of the frame shown in Fig. 29. Assume EI = 1 for all 3 members.

Ans. BF is a cantilever (as per first approximation)

$$k_{BA} = k_{BF} = \frac{EI}{h} = \frac{1}{1.6}$$

$$k_{BE} = \frac{3EI}{l/2} = \frac{3 \times 1}{2.4} = \frac{1}{0.8}$$

33. AB is a cantilever of span l, fixed at A. What is the carry over factor from B to A ?

Ans.

Fig. 30

Fig. 30 shows the situation. If we apply a moment M at B, it induces a moment of – M at the support A. Hence, carry over factor = – 1.

6

Deflection of Determinate Structures

1. Why it is necessary to compute deflections in structures ?

Ans. Computation of deflection of structures is necessary for the following reasons :

(*i*) If the deflection of a structure is more than the permissible, the structure will not look aesthetic and will cause psychological upsetting of the occupants.

(*ii*) Excessive deflection may cause cracking in the materials attached to the structure. For example, if the deflection of a floor beam is excessive, the floor finishes and partition walls supported on the beam may get cracked and unserviceable.

2. What is meant by 'cambering technique' in structures ?

Ans. Cambering is a technique applied on site, in which a slight upward curve is made in the structure/beam during construction, so that it will straighten out and attain the straight shape during loading. This will considerably reduce the downward deflection that may occur at later stages.

3. Name any four methods used for the computation of deflections in structures.

Ans. Methods used for computation of deflections are :

1. Double integration method
2. Macaulay's method
3. Conjugate beam method
4. Moment area method
5. Method of elastic weights
6. Virtual work method–Dummy unit load method
7. Strain energy method
8. Williot Mohr diagram method.

4. State the principle of virtual work. Mention its uses.

Ans. (*i*) According to the principle of virtual work,

Total external virtual work = Total internal virtual work

(*ii*) Σ (External forces × conjugate displacements)

= Σ (Internal forces × conjugate deformations)

$$\Sigma (F_{ext}\ \delta_{ext}) = \Sigma (F_{int} \times \delta_{int})$$

Use : Used to find unknown forces or displacements in loaded elastic structures.

5. Differentiate between the methods of virtual forces and virtual displacements.

Ans.

Method of virtual forces	*Method of virtual displacements*
Virtual force is applied in the calculation of real displacement.	Virtual displacements are imposed on a structure while the structure is subjected to real loadings.
Method can be used to determine the displacement in a structure.	Method can be used to determine the force on or in a structure.

6. State the law of conservation of energy.

Ans. The law of conservation of energy states that "If a conservative system is in equilibrium, the strain energy stored is equal to the work performed by the external loads."

7. List the methods used for the determination of deflections of joints in trusses.

Ans. The methods are:

1. Principle of virtual work (also called the unit-load method)
2. The angle-weights method
3. The joint-displacement equation method
4. The graphical method (Williot Mohr Diagram).

8. State the difference between strain energy method and unit load method in the determination of deflection of structures.

Ans. In strain energy method, an imaginary load P is applied at the point where the deflection is desired to be determined. P is equated to zero in the final step and the deflection is obtained.

In unit load method, an unit load (instead of P) is applied at the point where the deflection is desired.

9. What are the assumptions made in the unit load method ?

Ans. Assumptions made in unit load method are

1. The external and internal forces are in equilibrium
2. Supports are rigid and no movement is possible.
3. The material is strained well within elastic limit.

10. Give the equation that is used for the determination of deflection at a given point i in beams and frames.

Ans. Deflection at a point i is given by,

$$\delta i = \int_0^l \frac{M_x m_x \, dx}{EI}$$

where M_x = Moment at a section X due to the applied loads

m_x = Moment at a section X due to a unit load applied at the point i and in the direction of the desired displacement

EI = Flexural rigidity.

11. A simply supported beam AB of span 10 m is subjected to a concentrated load of 50 kN at a section C located 4 m from A. The deflection at the mid-span D is required. How many segments will the beam have to be divided into, if we use unit load method ? Sketch the same.

Ans.

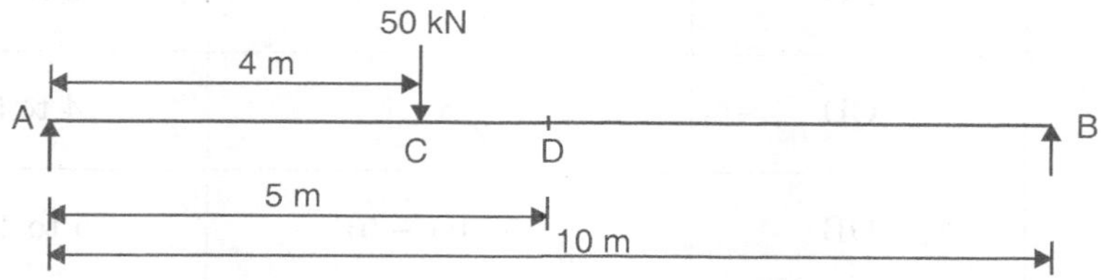

Fig. 1

The beam will have to be divided into three segments AC, CD and DB. Unit load will be applied at D.

12. For the beam in question 11, write down the equations for moments due to the external load, for the sections considered.

Ans.

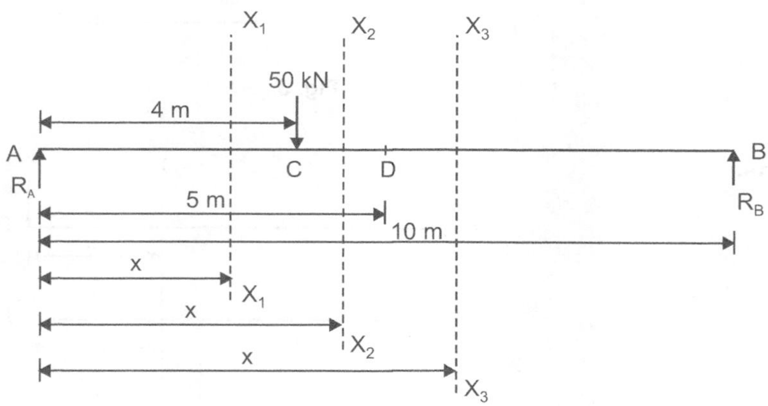

Fig. 2

Portion	M_x	*Limits*
AC	$R_A\, x$	0 to 4
CD	$R_A\, x - 50(x - 4)$	4 to 5
DB	$R_A x - 50(x - 4)$	5 to 10

13. For the beam in question 11, write down the equations for moments due to a unit load at D, for the sections considered.

Ans.

Portion	M_x	*Limits*
AC	$\frac{1}{2}x$	0 to 4
CD	$\frac{1}{2}x$	4 to 5
DB	$\frac{1}{2}x - 1(x-5)$	5 to 10

14. Briefly outline the steps for determining the rotation at the free end of the cantilever loaded as shown in Fig. 3.

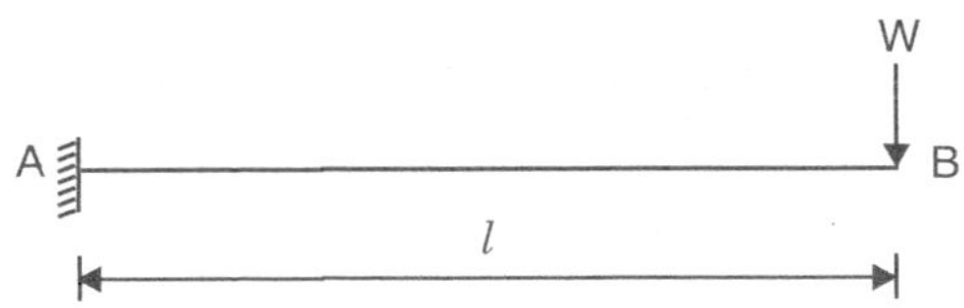

Fig. 3

Ans.

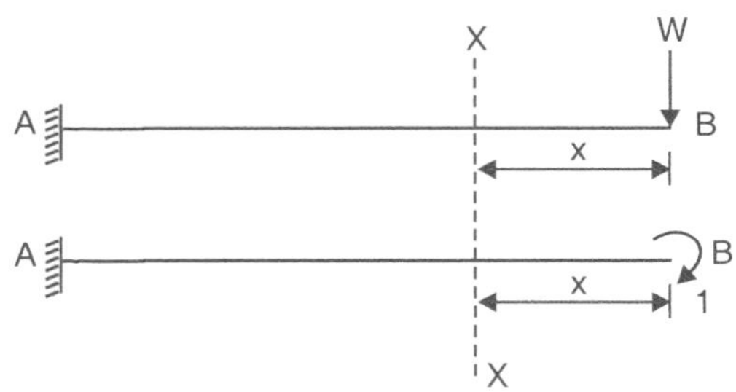

Fig. 4

1. $M_x = -Wx$
2. $m_x = -1$
3. $\theta_B = \int_0^l \frac{M_x\, m_x\, dx}{EI} = \int_0^l \frac{(-Wx)(-1)\, dx}{EI}$.

15. The horizontal displacement of the end D of the portal frame is required. Determine the relevant equations due to the unit load at appropriate point.

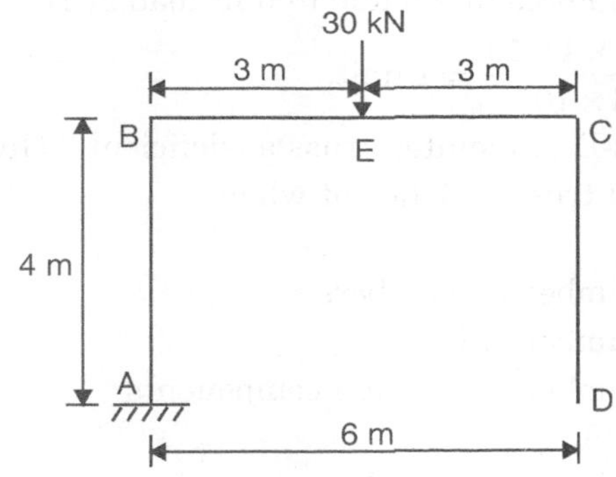

Fig. 5

Ans.

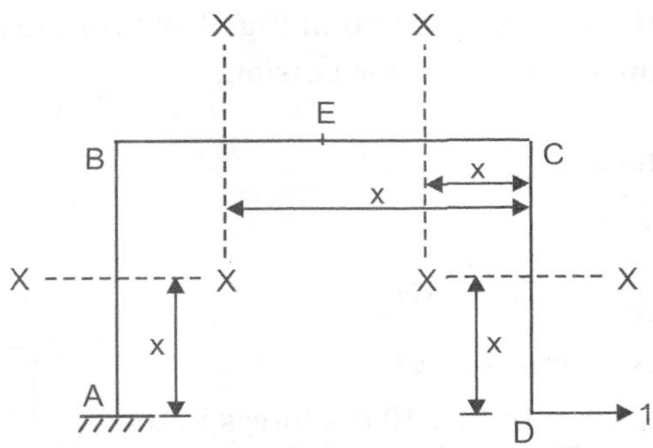

Fig. 6

Apply unit force in the horizontal direction at D. m_x values are tabulated as below :

Portion	m_x	*Limits*
DC	$1x$	0 to 4 m
CE	1×4	0 to 3 m
EB	1×4	3 to 6 m
BA	$1x$	0 to 4 m

16. Find the elongation of the tie in Fig. 7 by unit load method.

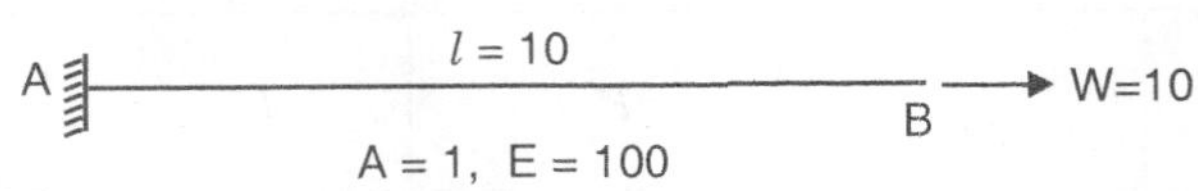

Fig. 7

Ans. Elongation $= \Delta_B = \dfrac{Pul}{AE}$

$P = 10$, Force in AB due to W

$u = 1$, Force in AB due to unit load at B

$$\Delta_B = \frac{10 \times 1 \times 10}{1 \times 100} = 1 \text{ unit.}$$

17. When will you call a 2-D pin jointed truss as deficient ? Give an example.

Ans. A 2-D pin jointed truss is deficient when,

$m < (2j - r)$

m = number of members

j = number of joints

r = number of reaction components

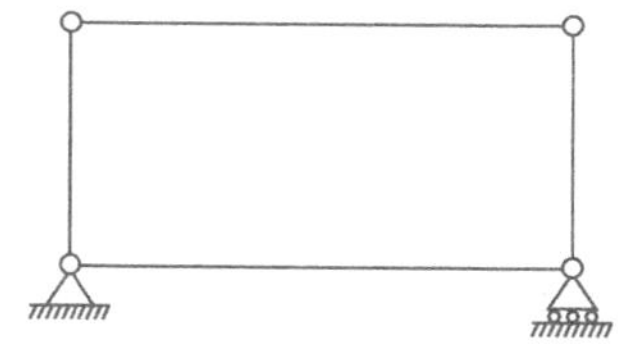

Fig. 8

In the example in Fig. 8,

$m = 4, j = 4; r = 3$ and

$m < (2j - 3)$; Hence deficient.

18. In a pin jointed truss, at the joint A shown in Fig. 9, obtain the force in the member AF.

Ans. Assuming the members to be under tension,

$\Sigma V = 0$ gives

$$F_{AF} \cos 45° + 50 = 0$$
$$(\uparrow) \qquad (\uparrow)$$

$$F_{AF} = -\frac{50}{\cos 45°} = -70.71 \text{ kN}$$

$\therefore$ $F_{AF} = 70.71$ kN (compressive)

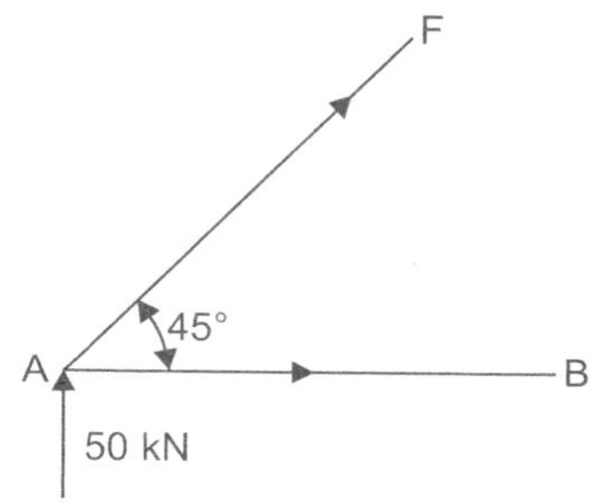

Fig. 9

19. Due to a load at B in the truss in Fig. 10 the forces in the members are as under. Determine the horizontal displacement of B by unit load method.

Member	$\frac{l}{AE}$	*Force due to P*
AB	0.01	80
BC	0.01	– 80
AC	0.01	40

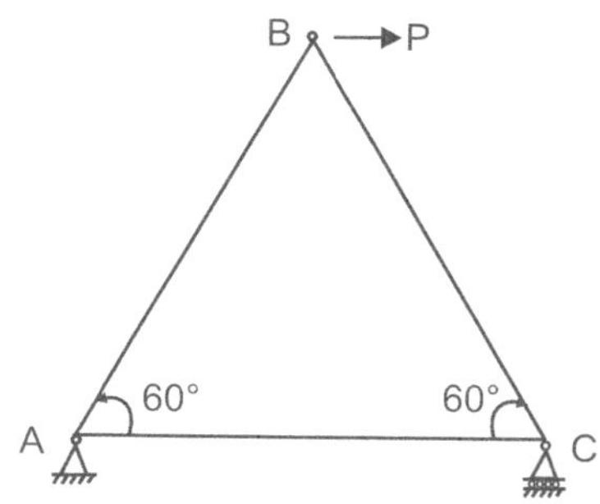

Fig. 10

Ans. $$\Delta_B = \sum \frac{Ful}{AE}$$

Member	$\frac{l}{AE}$	*F*	*U*	$\frac{Ful}{AE}$
AB	0.01	80	1	0.8
BC	0.01	– 80	– 1	0.8
AC	0.01	40	0.5	0.2
				1.8

$$\Delta_B = \sum \frac{Ful}{AE} = 1.8 \text{ units.}$$

20. Determine the rotation of the semi-circular beam in Fig. 11 due to a moment M_o, by unit load method.

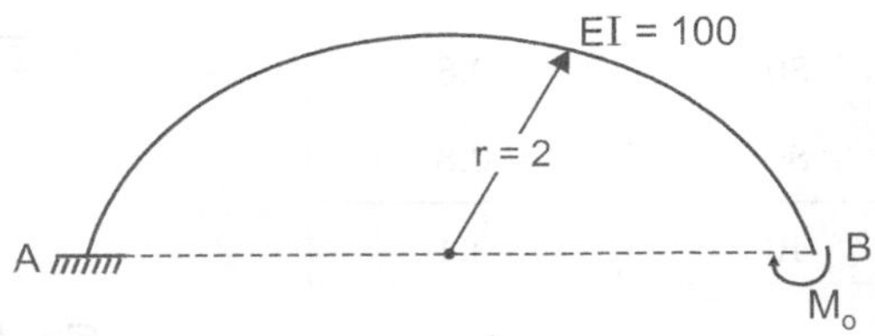

Fig. 11

Ans.

$$\theta_B = \int \frac{Mmds}{EI}$$

$m = 1$; $M = M_o$ EI = 100

$$\theta_B = \frac{M_o}{100}\int_0^{\pi} ds = \frac{M_o}{100}\pi r = \frac{2M_o}{100}\pi.$$

21. What is Williot diagram ?

Ans. Williot diagram is an ingenious graphical method of determining displaced positions of joints in a pin jointed frame/truss.

In this diagram the origin represents the undisplaced position of all joints.

For example, if a 2 member truss with 3 joints [Fig. 12(*a*)] has member AB stretching by 1.0 mm, and member BC shortening by 0.6 mm, the net displacement of B is found as in Fig. 12(*b*), the Williot diagram.

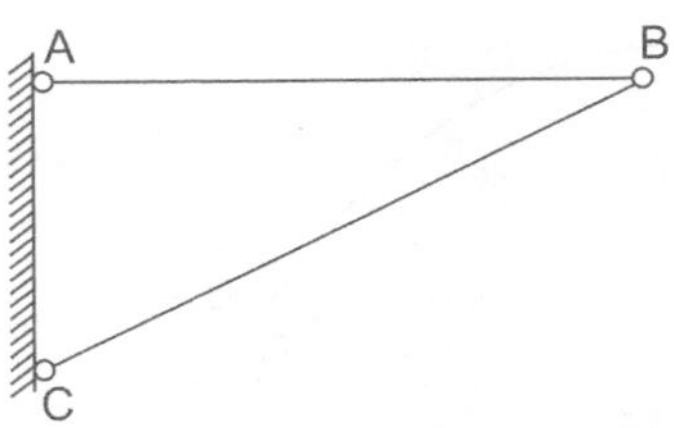

Fig. 12(a) Truss

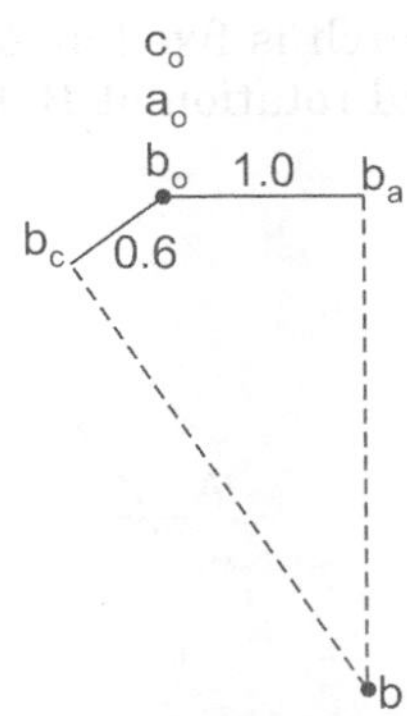

Fig. 12(b) Williot diagram

Along AB, B has to move out by 1.0 mm.

Along BC, B has to move towards C by 0.6 mm.

Since A is hinged, B can swing normal to BA.

Since C is hinged, B can swing normal to BC.

In the Williot diagram, $c_oa_ob_o$ is the origin (undisplaced positions). b_ob_a shows the extension of AB. b_ab shows the swing normal to BA. b_ob_c is the shortening of BC. b_cb is the swing normal to BC. b is the net displaced position of B.

22. Sketch the Williot's diagram for the truss in Fig. 12(c) to find Δ_B.

Ans.

Member	*F*	$\frac{Fl}{AE}$
AB	80	0.8
BC	– 80	– 0.8
AC	40	0.4

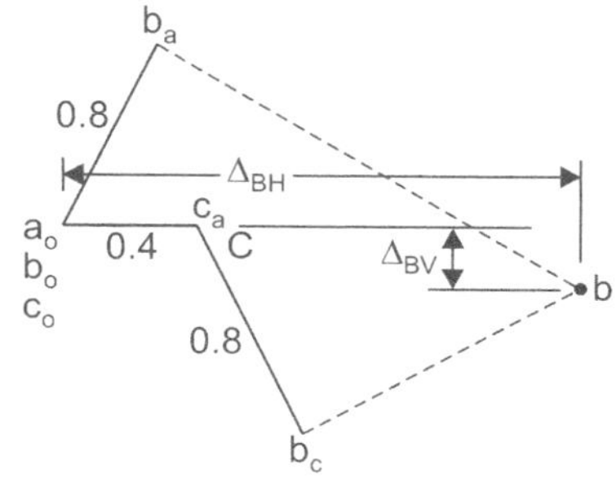

Fig. 12(c) Williot diagram

23. Find the rotation of end B of the beam shown in Fig. 13 by using Castigliano's I theorem.

Ans.

$$M = M_o$$

$$U = \int \frac{M^2\, dx}{2EI}$$

$$\theta_B = \frac{\partial U}{\partial M_o} = \frac{d}{dM_o} \int_o^l \frac{{M_o}^2\, dx}{2EI}$$

$$\theta_B = \frac{2M_o l}{2EI} = \frac{M_o l}{EI}.$$

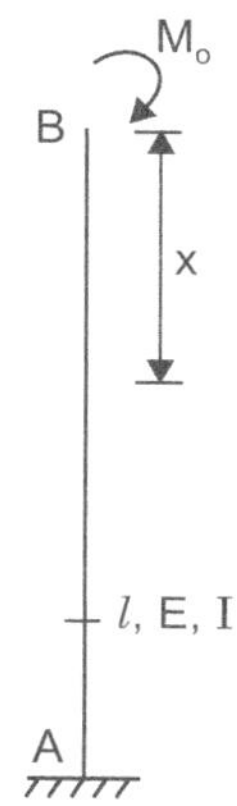

Fig. 13

24. In the above example, find the change of curvature of the beam AB.

Ans. Change of curvature $= \dfrac{\text{Change of slope}}{\text{Length}} = \dfrac{M_o l}{EI} \div l = \dfrac{M_o}{EI}.$

25. A parabolic arch is fixed at A and free at B (Fig. 14). Find expressions for the vertical deflection and rotation at B due to a moment M_o at B.

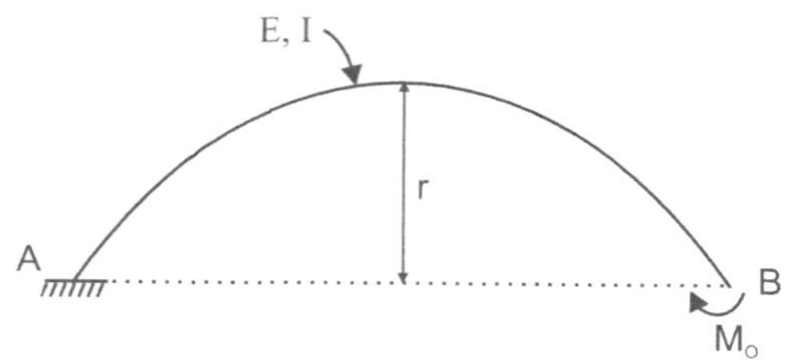

Fig. 14

Ans.

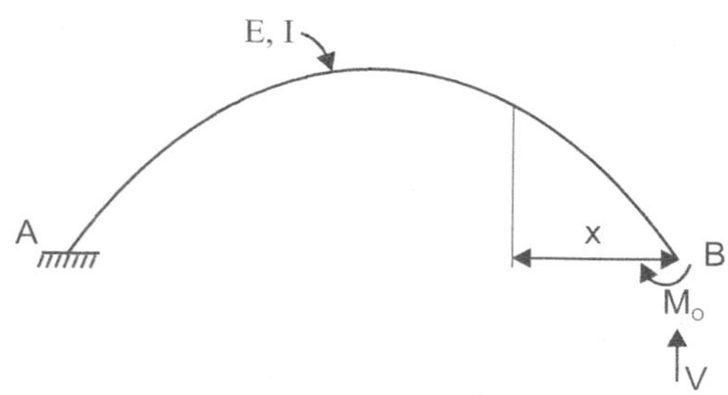

Fig. 15

Strain energy, $$U = \int \frac{M^2\, ds}{2EI}$$

Let us assume a virtual force V at B (Fig. 15).

$$M_x = Vx - M_o$$

$$\theta_B = \frac{\partial}{\partial M_o} U = \int_L \frac{M \dfrac{\partial M}{\partial M_o} ds}{EI}$$

$$\frac{\partial M}{\partial M_o} = -1$$

$$\theta_B = \int_L \frac{(Vx - M_o)(-1)ds}{EI}$$

$$V = 0$$

$$\theta_B = \frac{M_o L}{EI}, \text{ where L is the length of the arch}$$

$$\Delta_B = \frac{\partial U}{\partial V} \int_L \frac{(Vx - M_o)x ds}{EI}$$

$$V = 0$$

$$\therefore \quad \Delta_B = -\int_L \frac{M_o x ds}{EI}$$

This integration is difficult to perform.

26. Find the deflection of node C of the truss shown in Fig. 16. AE is the same for both members and equal to 40000 kN.

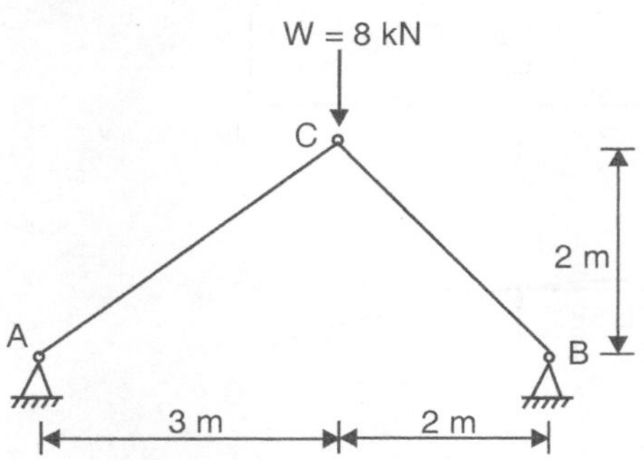

Fig. 16

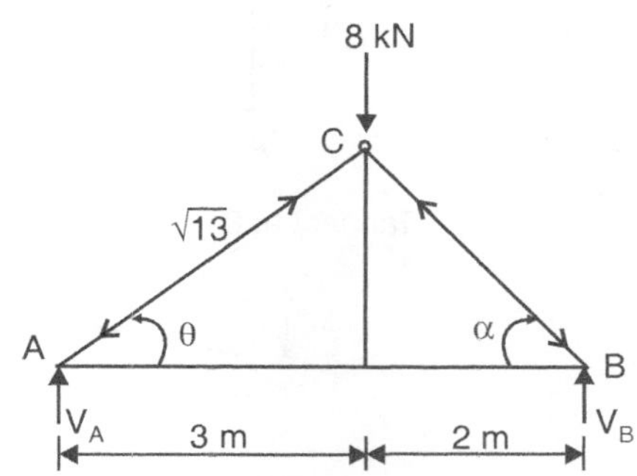

Fig. 17

Ans.

$$V_A = \frac{8 \times 2}{5} = 3.2 \text{ kN}$$

$$V_B = 4.8 \text{ kN}$$

$$F_{AC} = \frac{V_A}{\sin\theta} = \frac{3.2}{2}\sqrt{13} = 5.77$$

$$F_{BC} = \frac{V_A}{\sin\alpha} = \frac{4.8}{0.707} = 6.79$$

$$\Delta_C = \sum \frac{Ful}{AE}$$

Member	*l (m)*	*AE (kN)*	*F (kN)*	*u (= F ÷ 8)*	$\frac{Ful}{AE}$ *(m)*
AC	3.606	4×10^4	5.77	0.721	0.375×10^{-3}
BC	$2\sqrt{2}$	4×10^4	6.79	0.849	0.2037×10^{-3}

$$\Delta_C = 0.5788 \times 10^{-3} \text{ m}$$
$$\Delta_C = 0.5788 \text{ mm.}$$

27.

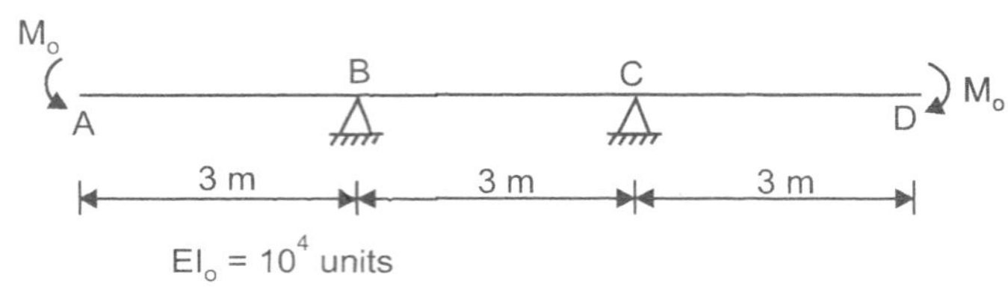

Fig. 18

$M_o = 20$ kNm.

Find the deflection of A in the beam shown in Fig. 18 by using Mohr's theorem.

Ans.

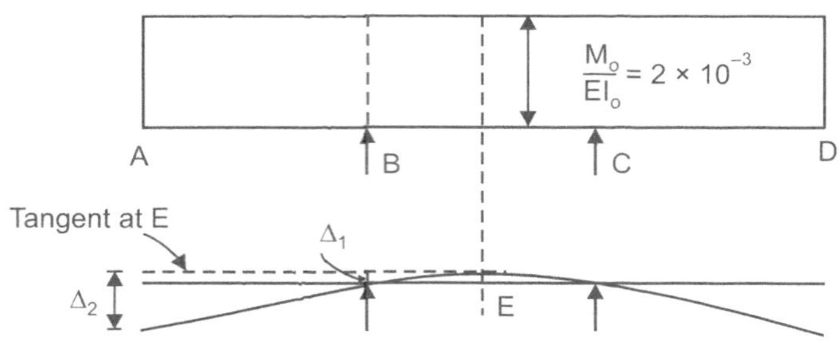

Fig. 19

Slope at mid-span E = 0

Δ_1 = intercept at B of tangents at B and E

= moment about B of area of $\frac{M}{EI}$ diagram between B and E

$$= \frac{M_o}{EI_o} \times 1.5 \times 0.75 = 2.25 \times 10^{-3}$$

$$\Delta_2 = \frac{M_o}{EI_o} \times 4.5 \times 2.25 \times 10^{-3} = 20.25 \times 10^{-3}$$

Deflection of A = $\Delta_2 - \Delta_1 = 18 \times 10^{-3}$ units.

28. Find the deflection at A of the beam shown in Fig. 20 by conjugate beam method.

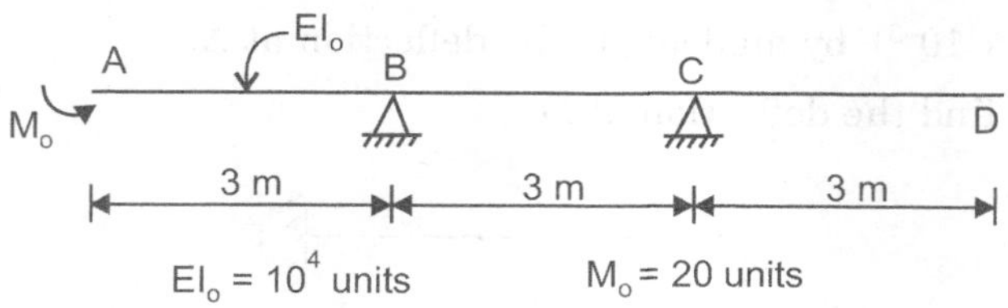

Fig. 20

Ans. The conjugate beam is

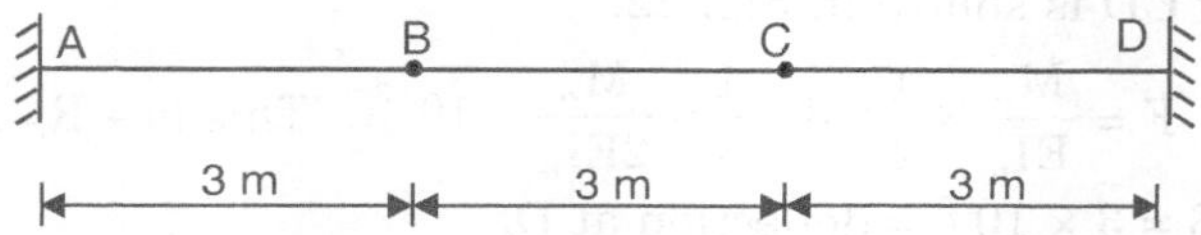

Fig. 21 (a)

The loading on the conjugate beam is the $\frac{M}{EI}$ diagram and is given by

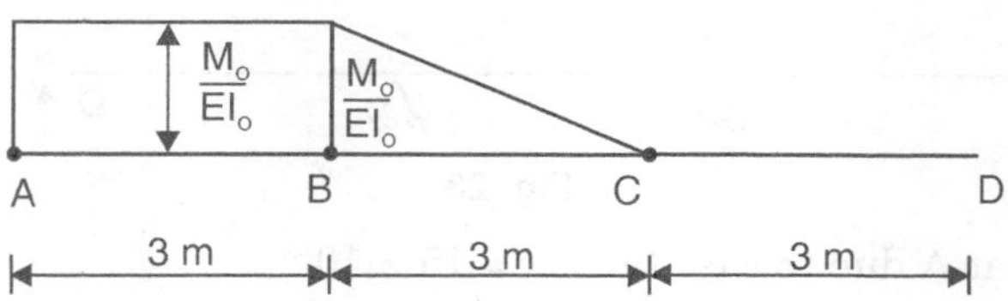

Fig. 21 (b)

The cantilever AB will have a udl of $\frac{M_o}{EI_o}$ plus downward reaction due to the triangular load on span BC.

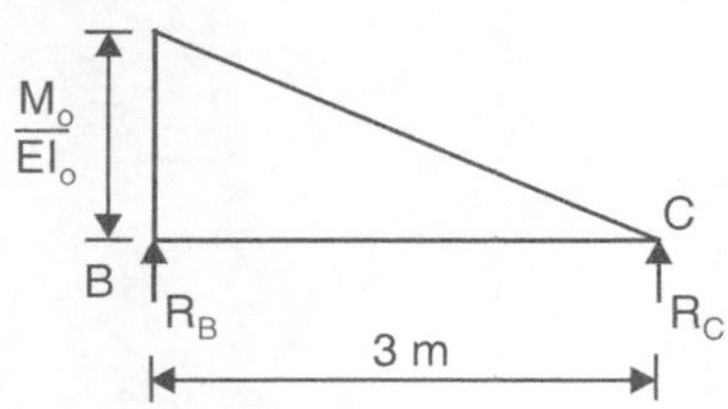

Fig. 21(c)

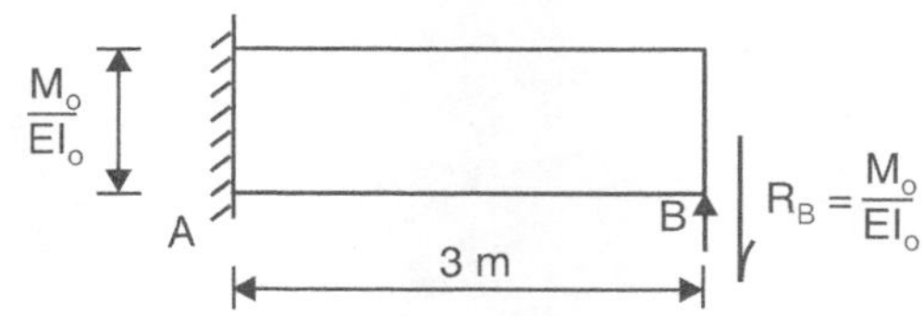

Fig. 21(d)

$$R_B = \frac{1}{2}\frac{M_o}{EI_o} \times 3 \times \frac{2}{3} = \frac{M_o}{EI_o}$$

$$\text{BM at A} = \frac{M_o}{EI_o} \times 3 \times \frac{3}{2} + R_B \times 3$$

$$= \frac{M_o}{EI_o} \times \left(\frac{9}{2} + 3\right) = 7.5\frac{M_o}{EI_o} = 15 \times 10^{-3}$$

This value (15×10^{-3}), by analogy, is the deflection at A.

29. In problem 28, find the deflection at D.

Ans.

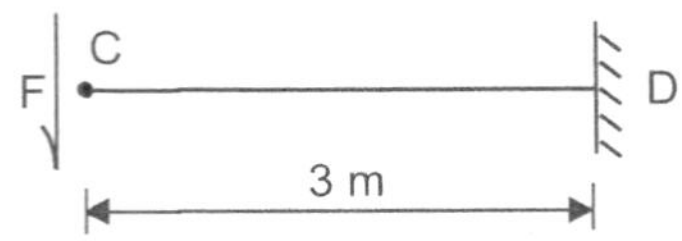

Fig. 22

The free body of CD is shown in Fig. 22.

$$F = \frac{M_o}{EI_o} \times \frac{1}{2} \times 3 \times \frac{1}{3} = \frac{M_o}{2EI_o} = 10^{-3} \quad \text{[This is } - R_C \text{ from Fig. 21 } (c)]$$

BM at D = F × 3 = 3 × 10^{-3} = deflection at D.

30. Use the results of questions 28 and 29 to solve problem 27.

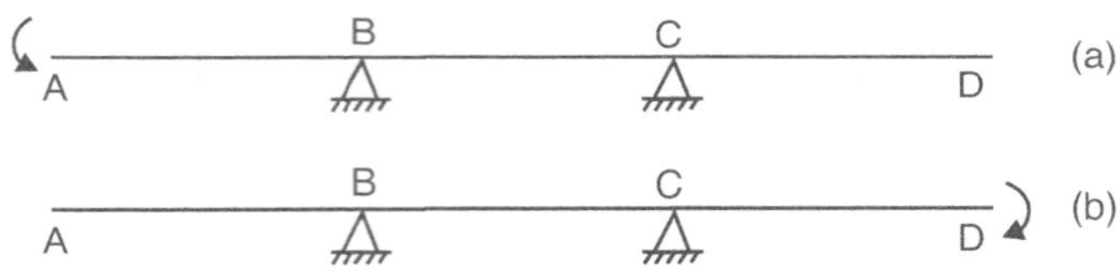

Fig. 23

Ans. Deflection at A due to case (*a*) = 15 × 10^{-3}

Deflection at A due to case (*b*) = 3 × 10^{-3}

Adding the above values, deflection at A due to loading in Fig. 18 = 18 × 10^{-3}.

7

Matrix Flexibility Method

1. What is meant by force method in structural analysis ?

Ans. A method in which the forces are treated as unknowns is known as force method. The following are force methods :

(*i*) Flexibility matrix method
(*ii*) Consistent deformation method
(*iii*) Claypeyron's 3 moment method
(*iv*) Column analogy method.

2. Define the term "flexibility" of a structure.

Ans. The term "flexibility"characterises the behaviour of a structure. It gives the displacement response to applied forces at the coordinates. This depends on the force displacement properties of the structure and the coordinate system adopted.

3. Define 'flexibility coefficient'.

Ans. Flexibility coefficient a_{ij} is defined as the displacement at coordinate i due to unit force at coordinate j in a structure. Flexibility coefficients makeup the elements of a flexibility matrix.

4. Why is flexibility method also called compatibility method or force method ?

Ans. Flexibility method begins with the superposition of forces and is hence known as force method.

Flexibility method leads to equations of displacement compatibility and is hence known as compatibility method.

5. A steel rod of length 'l', area of cross-section 'A' and Young's modulus 'E_s' is subjected to an axial unit load. What is its axial flexibility ?

Ans. Axial flexibility, $[a] = \dfrac{l}{AE_s}$.

6. The cantilever beam shown in Fig. 1 (*a*) is subjected to actions A_1 and A_2 at the free end. The corresponding displacements are denoted by D_1 and D_2. Write down its flexibility matrix.

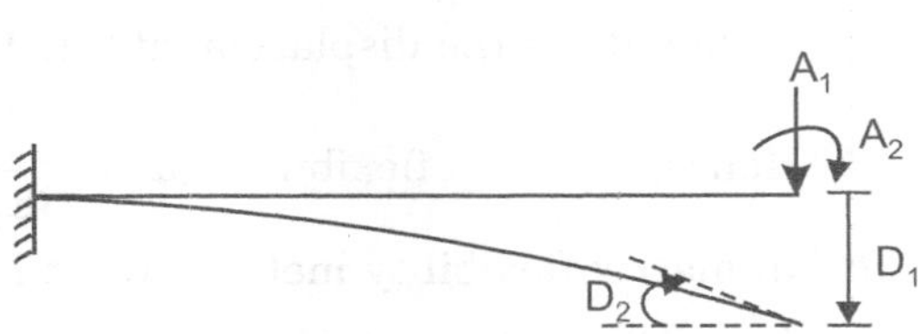

Fig. 1 (a)

Ans. Applying $A_1 = 1$

$$D_{11} = \left(\frac{1}{2} \times l \times \frac{l}{EI}\right) \times \frac{2}{3} l = \frac{l^3}{3EI}$$

$$D_{12} = \left(\frac{1}{2} \times l \times \frac{l}{EI}\right) = \frac{l^2}{2EI}$$

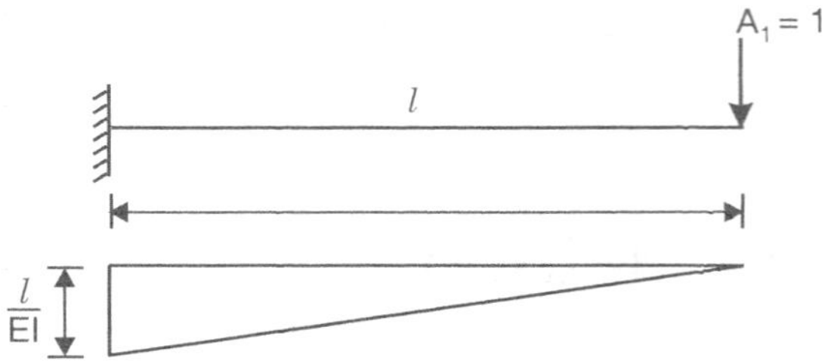

Fig. 1 (b)

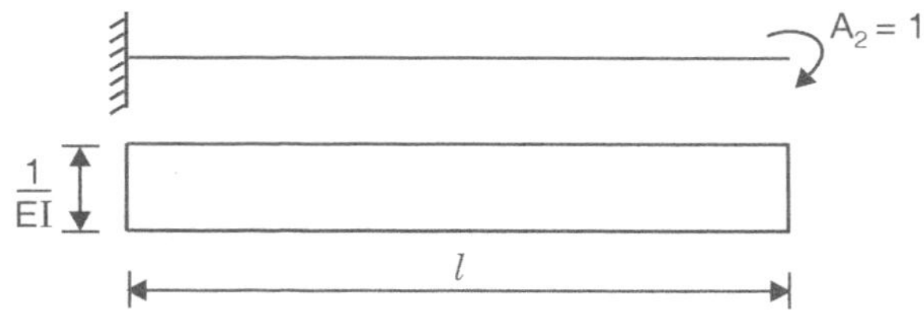

Fig. 1 (c)

Applying $A_2 = 1$

$$D_{12} = \left(1 \times \frac{l}{EI}\right) \frac{l}{2} = \frac{l^2}{2EI}$$

$$D_{22} = 1 \times \frac{l}{EI} = \frac{l}{EI}$$

Flexibility matrix, $$[a] = \begin{bmatrix} D_{11} & D_{12} \\ D_{21} & D_{22} \end{bmatrix} = \begin{bmatrix} \frac{l^3}{3EI} & \frac{l^2}{2EI} \\ \frac{l^2}{2EI} & \frac{l}{EI} \end{bmatrix}.$$

7. Write the flexibility matrix of the simply supported beam element in Fig. 2.

Fig. 2

Ans. Flexibility matrix $[a] = \frac{l}{6\,EI} \begin{bmatrix} 2 & -1 \\ -1 & 2 \end{bmatrix}$

8. A cantilever of span 'l' carries a load 'W' at the free end. Determine the flexibility of the beam.

Ans. Displacement at the free end due to W = $\frac{Wl^3}{3EI}$

Flexibility is the displacement due to unit force.

Hence, flexibility, $[a] = \frac{l^3}{3EI}$.

9. In matrix flexibility method, what is the compatibility equation ?

Ans. The matrix equation

$$\{u\} = [a]\,\{F\}$$

is sometimes called the compatibility equation. Herein, {u} are the system displacements, [a] the flexibility matrix and {F} the system co-ordinate forces.

{F} would include the indeterminate forces also, besides external forces.

10. A simple beam is subjected to unit clockwise moment at mid-span. Find the flexibility coefficient. Take EI = l = 1.

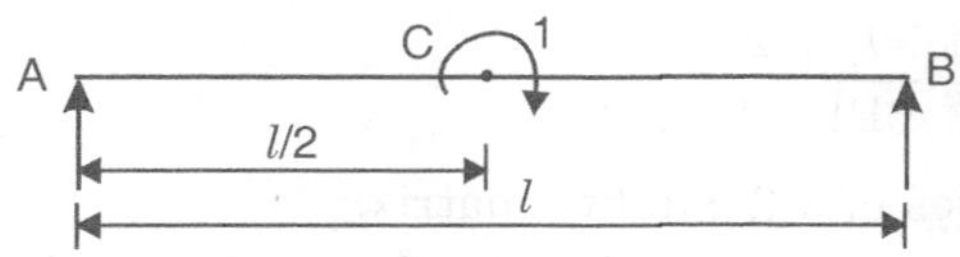

Fig. 3

Ans. This loading is anti-symmetrical. Hence C will not deflect up or down but will stay at the same level. It can be treated as a support with zero reaction. The moment will be shared by CA and CB as 0.5 and 0.5. The rotation of CA (= rotation of CB) = $\dfrac{\frac{1}{2}\, l/2}{3\text{EI}} = \dfrac{l}{12\text{EI}}$.

$\therefore$ Flexibility = rotation per unit moment = $\dfrac{l}{12\text{EI}}$.

11. Develop the flexibility matrix for the simply supported beam with reference to the coordinates shown in Fig. 4.

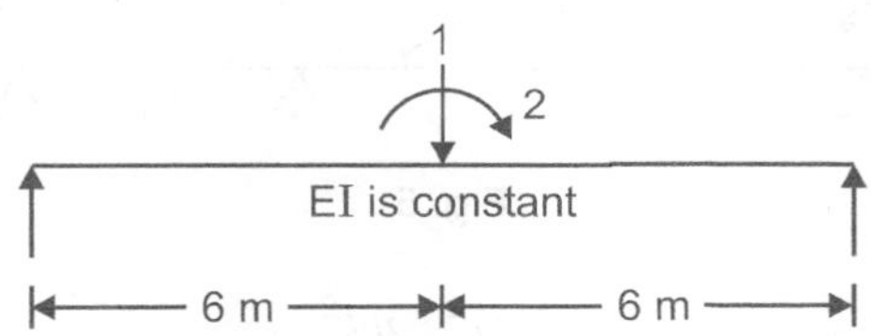

Fig. 4

Ans. Applying unit force along coordinate 1,

Displacement along 1, $a_{11} = \dfrac{Wl^3}{48\text{EI}} = \dfrac{1 \times 12^3}{48\text{EI}} = \dfrac{36}{\text{EI}}$

Displacement along 2, a_{21} = 0 (symmetry)

Applying unit force along coordinate 2, (Refer previous answer)

Displacement along 1, a_{12} = 0 (Deflection along 1)

Displacement along 2, $a_{22} = \dfrac{l}{12\text{EI}} = \dfrac{12}{12\text{EI}} = \dfrac{1}{\text{EI}}$

Hence, $[a] = \begin{bmatrix} \dfrac{36}{\text{EI}} & 0 \\ 0 & \dfrac{1}{\text{EI}} \end{bmatrix} = \dfrac{1}{\text{EI}}\begin{bmatrix} 36 & 0 \\ 0 & 1 \end{bmatrix}$

[Here 1 and 2 are called 'uncoupled']

12. Give the element flexibility of a truss member and of a beam element

Ans. Element flexibility of a truss member

$$[\alpha] = \frac{l}{AE}$$

Element flexibility of a beam element

$$[\alpha] = \frac{l}{6EI}\begin{bmatrix} 2 & -1 \\ -1 & 2 \end{bmatrix}$$

13. Mention the properties of a flexibility matrix.

Ans. The following are the properties of a flexibility matrix:

1. The flexibility matrix is a square matrix of order $[s_i \times s_i]$
   $[s_i \rightarrow$ Degree of static indeterminacy$]$
2. The flexibility matrix is a symmetrical matrix.
3. The leading diagonal elements of a flexibility matrix are always positive. Other elements may be positive or negative.

14. The system in Fig. 5 recognises co-ordinates 1, 2, 3 and 4 as marked. What forces shall be assigned to the coordinates due to a udl of w/m over a span as shown in Fig. 5

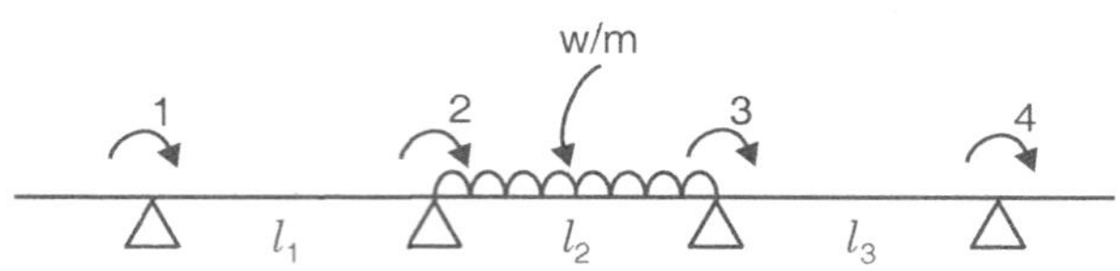

Fig. 5

Ans. FEM due to w at 2 and 3 are $-\frac{wl^2}{12}, \frac{wl^2}{12}$.

Forces to be assigned to the co-ordinates are

$$[F_1, F_2, F_3, F_4]^T = \begin{bmatrix} 0 & \frac{wl^2}{12} & -\frac{wl^2}{12} & 0 \end{bmatrix}^T.$$

15. What is a primary structure ?

Ans. In the analysis of statically indeterminate structures, the degree of static indeterminacy is determined and the structure is made statically determinate by releasing all the redundants. The released structure is known as the primary structure.

16. State the principle of superposition of forces.

Ans. The effect of applying several forces in a structure can be obtained by superposing the effects due to the individual forces, one at a time. This is the principle of superposition of forces.

This is valid for linear, elastic, conservative structures.

17. Define the force transformation matrix.

Ans. The connectivity matrix which relates the element forces {P} and the system forces {F} is known as the force transformation matrix [*b*].

$$\{P\} = [b]\{F\}$$

18. Will it be possible to generate the flexibility matrix for an unstable structure ?

Ans. No. It is not possible because a structure has to be stable for generating the flexibility matrix.

19. Write the flexibility matrix for coordinates 1 and 2 of the beam shown in Fig. 6.

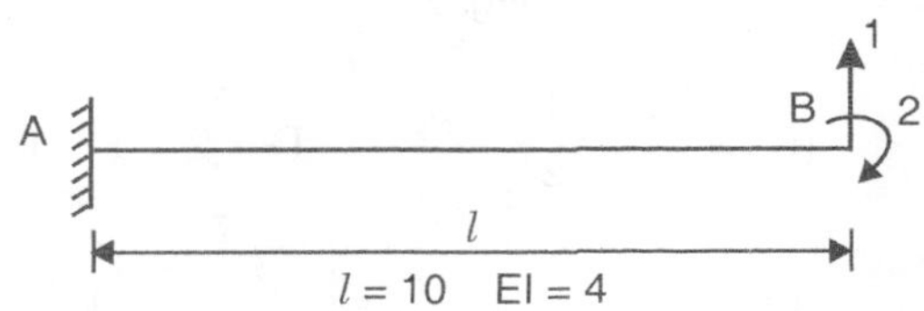

Fig. 6

Ans.

$$[a] = \frac{l}{EI}\begin{bmatrix} \frac{l^2}{3} & -\frac{l}{2} \\ -\frac{l}{2} & 1 \end{bmatrix}$$

$$= \frac{10}{4}\begin{bmatrix} \frac{100}{3} & -\frac{l}{2} \\ -\frac{10}{2} & 1 \end{bmatrix} = 2.5\begin{bmatrix} 33.3 & -5 \\ -5 & 1 \end{bmatrix}$$

20. What is the relation between the flexibility matrix and stiffness matrix ?

Ans. The relation between the flexibility matrix and stiffness matrix is that, one is the inverse of the other, when they both exist.

21. State any two methods of matrix inversion.

Ans. Methods of matrix inversion are

(*i*) Adjoint method

(*ii*) The Gauss-jordon method (By linear transformation)

(*iii*) The Choleski method (By factorisation)

(*iv*) Partitioning method.

22. What is a transformation matrix ?

Ans. In the analysis, for convenience we fix element coordinates coincident with member axes called local coordinates. It is normally necessary to define a coordinate system dealing with the entire structure, called system coordinates or global coordinates. It becomes necessary to transform information from member coordinates to structure coordinates and vice versa. The matrix used for this transformation is known as transformation matrix.

23. Find the slope at the prop end of a prismatic propped cantilever beam, given that the stiffness matrix of the structure as shown in Fig. 7 is given by $[K] = \begin{bmatrix} 1 & 1 \\ 1 & 2 \end{bmatrix}$ and moment at the prop is 2.

Fig. 7

Ans.

$$[K] = \begin{bmatrix} 1 & 1 \\ 1 & 2 \end{bmatrix} \qquad D = 2 - 1 = 1$$

$$[a] = \frac{1}{D}\begin{bmatrix} 2 & -1 \\ -1 & 1 \end{bmatrix} = \begin{bmatrix} 2 & -1 \\ -1 & 1 \end{bmatrix}$$

$$\{\delta\} = \begin{Bmatrix} \delta_1 \\ \delta_2 \end{Bmatrix} = [a]\{P\} = \begin{bmatrix} 2 & -1 \\ -1 & 1 \end{bmatrix}\begin{Bmatrix} P_1 \\ P_2 \end{Bmatrix} \qquad ...(1)$$

$\delta_1 = 0$ $\qquad$ δ_2 = rotation at B (to be determined)

P_1 is unknown ; $P_2 = 2$.

From (1), $\qquad \delta_1 = 0 = 2P_1 - P_2 = 2P_1 - 2$; $P_1 = 1$

$\qquad \delta_2 = -P_1 + P_2 = -1 + 2 = 1$

The rotation at B is 1 unit.

24. Prove the symmetry property of the flexibility and stiffness matrices.

Ans. Let us take 2 terms in a stiffness matrix.

$$K_{ij} \text{ and } K_{ji}$$

K_{ij} is the force required at i to cause unit displacement at j. Hence $\frac{1}{K_{ij}}$ is the displacement at j due to a unit load at i.

By Maxwell's law of reciprocal displacement,

$$\frac{1}{K_{ij}} = \frac{1}{K_{ji}}; \quad \text{hence } K_{ij} = K_{ji}.$$

The [K] matrix is therefore symmetrical.

By a similar argument the $[a]$ matrix also is symmetrical.

25. A strut ABC is held in position at an intermediate point B (Fig. 8). Find its 2 × 2 flexibility matrix.

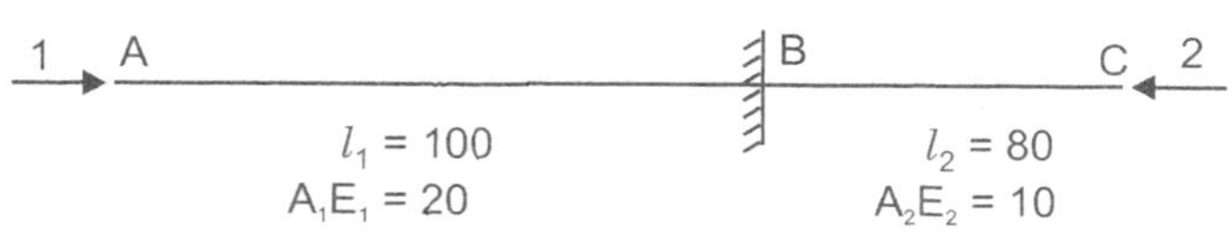

Fig. 8

Ans. $$[a] = \begin{bmatrix} \dfrac{l_1}{A_1E_1} & 0 \\ 0 & \dfrac{l_2}{A_2E_2} \end{bmatrix} = \begin{bmatrix} \dfrac{100}{20} & 0 \\ 0 & \dfrac{80}{10} \end{bmatrix} = \begin{bmatrix} 5 & 0 \\ 0 & 8 \end{bmatrix}$$

26. For the member shown in Fig. 8, find the stiffness matrix.

Ans. $$[K] = \begin{bmatrix} \dfrac{A_1E_1}{l_1} & 0 \\ 0 & \dfrac{A_2E_2}{l_2} \end{bmatrix} = \begin{bmatrix} \dfrac{1}{5} & 0 \\ 0 & \dfrac{1}{8} \end{bmatrix}.$$

27. Define local and global coordinates.

Ans. Local coordinates. Coordinates defined along the individual member axes locally.

Global coordinates. Common coordinate system dealing with the entire structure. Also known as system coordinates.

28. For which of these systems can we *not* form a flexibility matrix ?

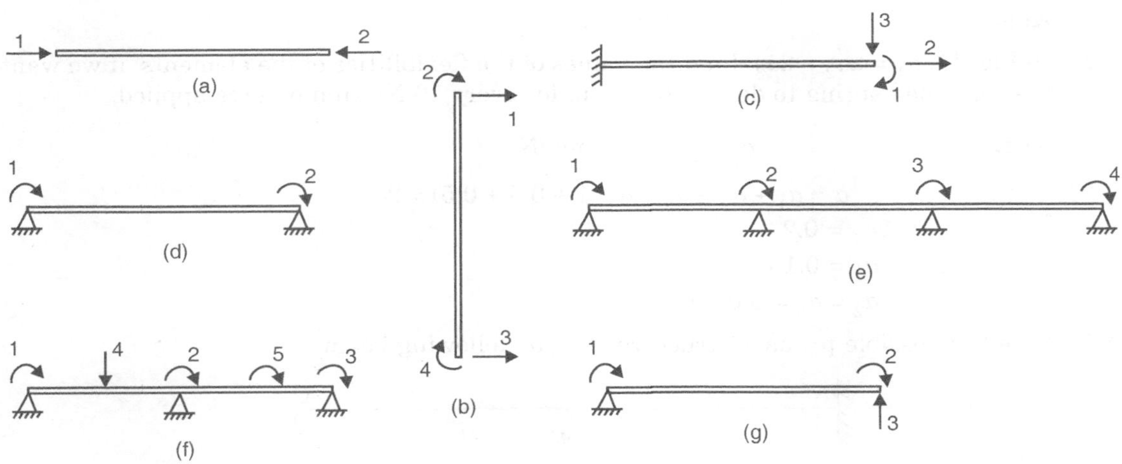

Fig. 9

Ans. (*a*), (*b*), (*g*).

29. The combined spring shown in Fig. 10 is madeup of 3 spring elements of flexibilities a_1, a_2 and a_3. What is the combined flexibility 'a' ?

Fig. 10

Ans. $$\frac{1}{a} = \frac{1}{a_1} + \frac{1}{a_2} + \frac{1}{a_3}$$

$$a = \frac{1}{\frac{1}{a_1} + \frac{1}{a_2} + \frac{1}{a_3}}.$$

30. In Fig. 10, if $a_1 = a_2 = 16$, find the value of a_3 so that the combined flexibility would be 4.

Ans.
$$\frac{1}{16} + \frac{1}{16} + \frac{1}{a_3} = \frac{1}{4}$$
$$\frac{1}{a_3} = 0.25 - 0.125 = 0.125$$
$$a_3 = 8.$$

31. In Fig. 11, the flexibilities of the 3 elements of the combined spring are a_1, a_2 and a_3. Determine the combined flexibility 'a'.

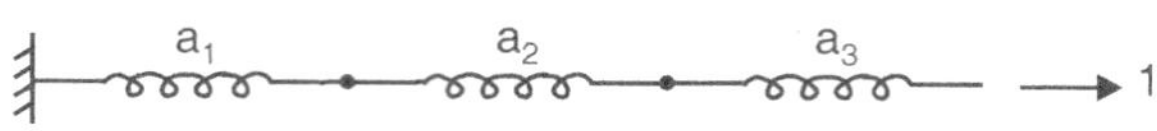

Fig. 11

Ans.
$$a = a_1 + a_2 + a_3.$$

32. In Fig. 11, $a_1 = 2a_2 = 2a_3$. Find the values of the flexibilities of the elements if we want the combined spring to deform by 2 mm for every 10 Newton of force applied.

Ans.
$$a = \frac{2}{10} = 0.2 \text{ mm/N}$$
$$a = a_1 + a_2 + a_3 = a_1\,(1 + 0.5 + 0.5) = 2a_1$$
$$2a_1 = 0.2$$
$$a_1 = 0.1 \text{ mm/N}$$
$$a_2 = a_3 = 0.05 \text{ mm/N}.$$

33. Give the possible primary structures for the following beam.

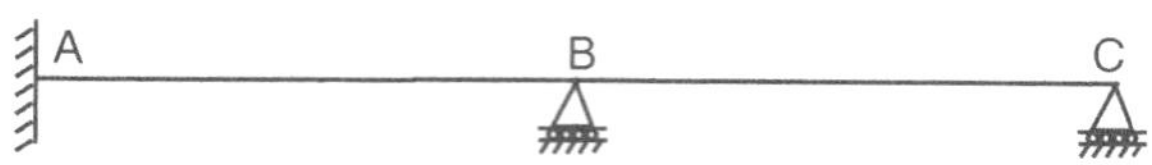

Fig. 12 (a)

Ans. Degrees of static indeterminacy, I = V.R. + E.M. – 2 (for vertical loading)

$$I = 3 + 1 - 2 = 2$$

Primary structures | Redundants

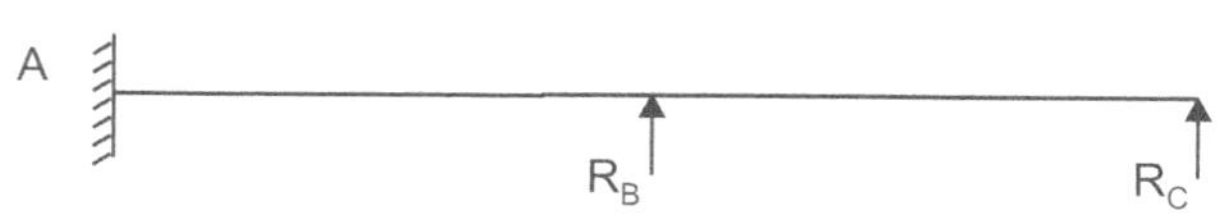

(b) — R_B and R_C

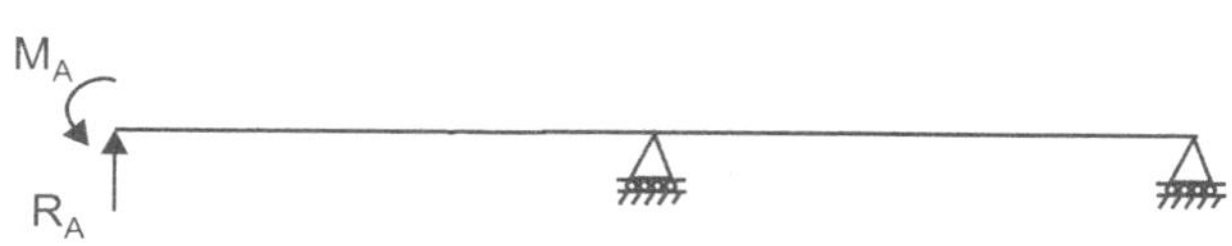

(c) — R_A and M_A

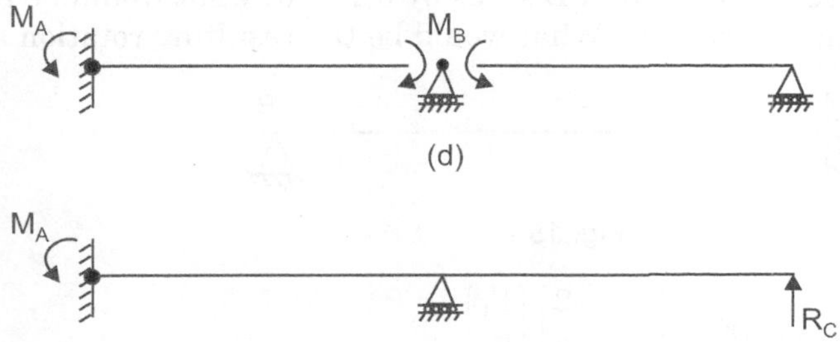

(d) M_A and M_B

(e) M_A and R_C

Fig. 12 (b) to (e)

34. Suggest primary structures for the following beam.

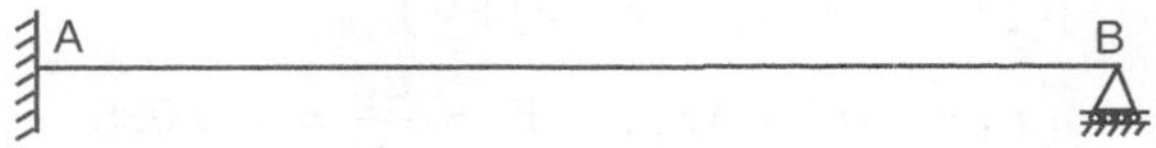

Fig. 13 (a)

Ans. Degree of static indeterminacy, I = 1

Primary structures	Redundants
(b)	R_B
(c)	M_A

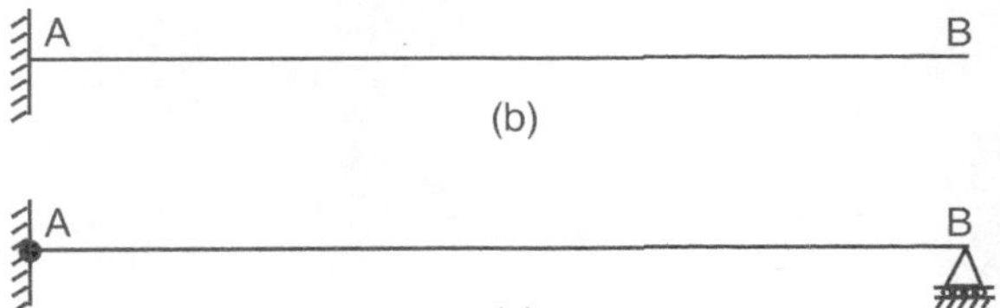

Fig. 13 (b) and (c)

35. The flexibility matrix of the structure shown in Fig. 14 is $[a] = \begin{bmatrix} 4 & 2 \\ 2 & 3 \end{bmatrix}$. Find its stiffness matrix.

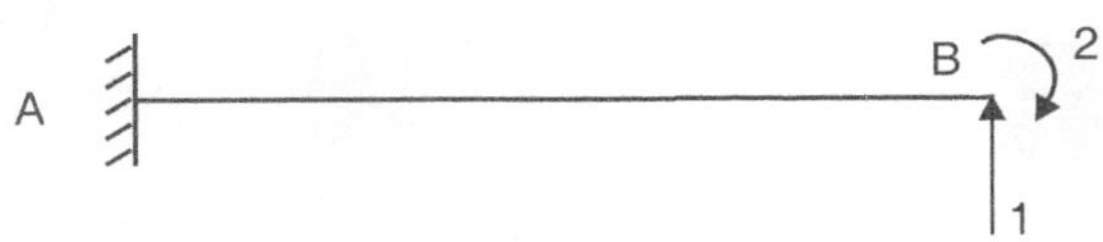

Fig. 14

Ans.

$$[K] = [a]^{-1} = \frac{1}{8}\begin{bmatrix} 3 & -2 \\ -2 & 4 \end{bmatrix}$$

36. In the beam shown in Fig. 15, if support B sinks by 0.1 unit, what would be reaction at B ? Use the $[a]$ matrix in question 35. What would be the resulting rotation at B ?

Fig. 15

Ans.

$$u = [a]\,\{F\} = \begin{bmatrix} 4 & 2 \\ 2 & 3 \end{bmatrix} \begin{Bmatrix} F_1 \\ F_2 \end{Bmatrix}$$

$u_1 = -\,0.1$ (sinking of support)

u_2 is unknown

F_2 is zero.

$$\begin{Bmatrix} u_1 \\ u_2 \end{Bmatrix} = [a] \begin{Bmatrix} F_1 \\ F_2 \end{Bmatrix} = \begin{bmatrix} 4 & 2 \\ 2 & 3 \end{bmatrix} \begin{Bmatrix} F_1 \\ 0 \end{Bmatrix}$$

$$u_1 = -\,0.1 = 4F_1; \quad F_1 = \frac{u_1}{4} = -\,0.025$$

$$u_2 = 2F_1 = -\,0.05 \text{ units.}$$

8

Matrix Stiffness Method

1. What is displacement method of analysis ?

Ans. A method of analysis in which the displacements are treated as the primary unknowns is known as displacement method of analysis.

2. Define the term "stiffness" of a structure.

Ans. The term "stiffness"characterises the behaviour of a structure. It gives the force required to produce the given displacements at the coordinates. This depends upon the force displacement properties of the structure and the coordinate system adopted.

3. Define "stiffness coefficient".

Ans. Stiffness coefficient k_{ij} is defined as the force developed at coordinate i due to unit displacement at coordinate j in a structure, with the condition that no displacement occurs at other co-ordinates, except at j. Stiffness coefficients make up the elements of a stiffness matrix.

4. What are the other names of the stiffness method ?

Ans. Stiffness method is also known as displacement method (or) equilibrium method.

5. Why is stiffness method also called displacement method or equilibrium method ?

Ans. Stiffness method begins with the superposition of displacements and is hence known as displacement method.

Stiffness method leads to equations of equilibrium and is hence known as equilibrium method.

6. Explain constrained displacements.

Ans.

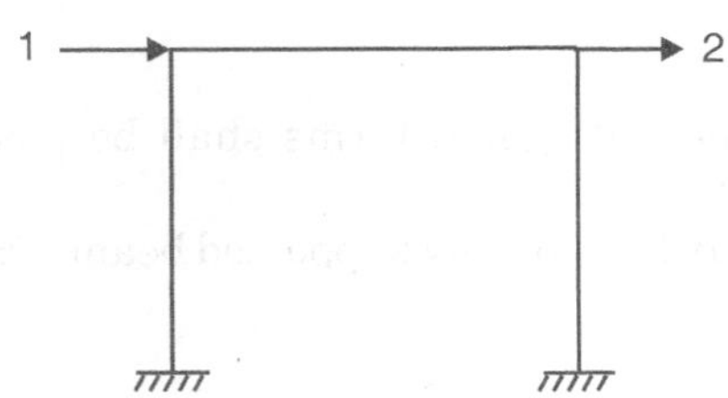

Fig. 1

Constrained displacements are displacements that cannot take place independently. *i.e.*, the displacement along coordinates 1 and 2 cannot take place independently. We can not apply a displacement at 1 and arrest the displacement at 2 (if we assume that there are no axial deformations). 1 and 2 are constrained displacements.

7. What is the size of the stiffness matrix when the kinematic indeterminacy is 4 ?

Ans. When the kinematic indeterminacy is 4, the size of the stiffness matrix is 4 × 4.

8. Which property of a structure determines the size of its stiffness matrix ?

Ans. The kinematic indeterminacy of a structure determines the size of its stiffness matrix.

Example: The kinematic indeterminacy of structure in Fig. 2 is 4 (θ_B, θ_C, θ_D and δ)

Hence the size of the (Global)

Stiffness matrix is (4 × 4)

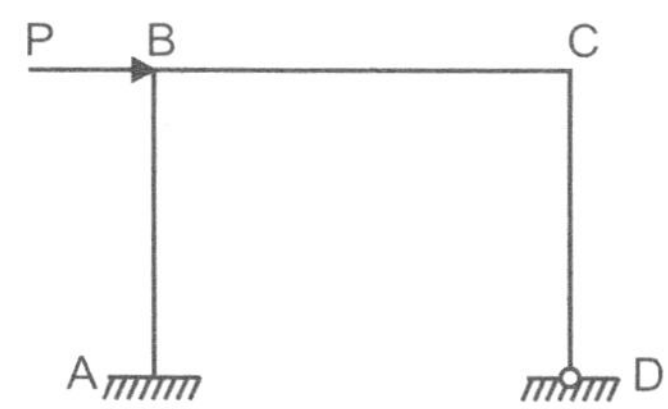

Fig. 2

9. Write down the equilibrium equation used in stiffness method.

Ans. Equilibrium equation used in stiffness method is $[K]\{u\} = \{f\}$

where [K] = matrix of stiffness coefficients

$\{u\}$ = vector of unknown displacements at nodes

$\{f\}$ = vector of nodal forces.

10. When do we prefer stiffness method ?

Ans. When the kinematic indeterminacy of a structure is less than the static indeterminacy, stiffness method is preferred.

11. Show that the stiffness matrix is symmetric.

Ans. A term k_{ij} in a stiffness matrix is equal to the force developed at co-ordinate i due to a unit displacement at j. Conversely k_{ji} is the force at j due to a unit displacement at i.

Maxwell's law of reciprocal deflection implies that the deflection at i due to a unit load at j is equal to the deflection at j due to the unit load at i. Accordingly, $\frac{1}{k_{ij}} = \frac{1}{k_{ji}}$. Hence, $k_{ij} = k_{ji}$.

Therefore, [K] matrix is symmetric.

12. What are the properties of a structure stiffness matrix ?

Ans. A valid stiffness matrix of a stable, linear, elastic structure has to be

(*i*) symmetric

(*ii*) positive definite

(*iii*) strong on the diagonal (diagonal terms shall be positive, non zero and dominant compared to off diagonal terms).

13. Write the stiffness matrix for a simply supported beam element with rotation at supports as coordinates.

Ans.

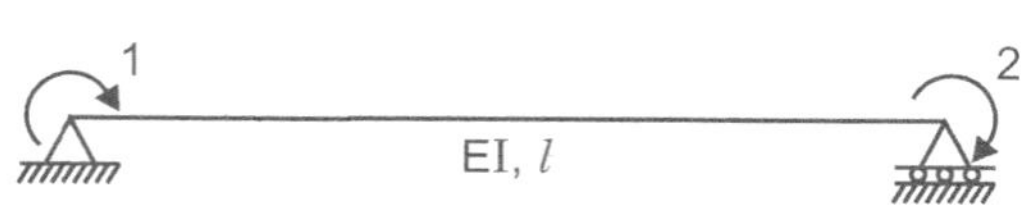

Fig. 3

Stiffness matrix, $[k] = \dfrac{EI}{l}\begin{bmatrix} 4 & 2 \\ 2 & 4 \end{bmatrix}$

14. Derive the element stiffness matrix for a plane truss member.

Ans.

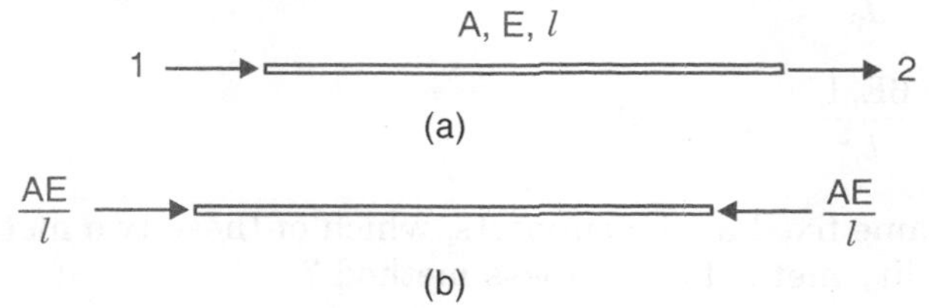

Fig. 4

The truss member shown in Fig. 4 (*a*) has 2 co-ordinates. The stiffness matrix [K] would be of size 2 × 2. The first column of [K] would be obtained by applying a unit displacement along 1 and arresting displacement at 2. The forces needed would be as in Fig. 4 (*b*). These 2 values make column 1 of [K] matrix and are written as $\dfrac{AE}{l}[1 \quad -1]^T$. Like wise the forces needed at 1 and 2 to cause a unit displacement along 2 only and keep the displacement at 1 as zero are $\dfrac{AE}{l}[-1 \;\; 1]$. Hence, $[K] = \dfrac{AE}{l}\begin{bmatrix} 1 & -1 \\ -1 & 1 \end{bmatrix}$.

15.

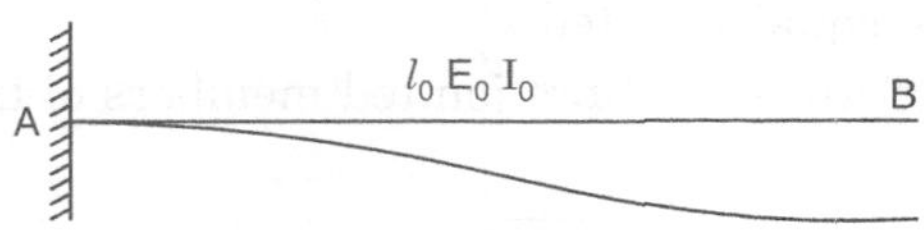

Fig. 5

The free end B of the cantilever shown in Fig. 5 undergoes a unit vertical displacement without any rotation. What is the moment induced at A ?

Ans. $M_A = \dfrac{6E_0I_0}{l_0^{\,2}}$

16. In the beam in question 15 what are the forces/moments required at B to sustain this displacement.

Ans. $M_B = \dfrac{6E_0I_0}{l_0^{\,2}}$ anti-clockwise

$V_B = \dfrac{12E_0I_0}{l_0^{\,3}}$ downwards.

17. The free end B of the beam shown in Fig. 6 undergoes a unit-clockwise rotation with no vertical displacement. What are the forces at A and B, consistent with this displacement?

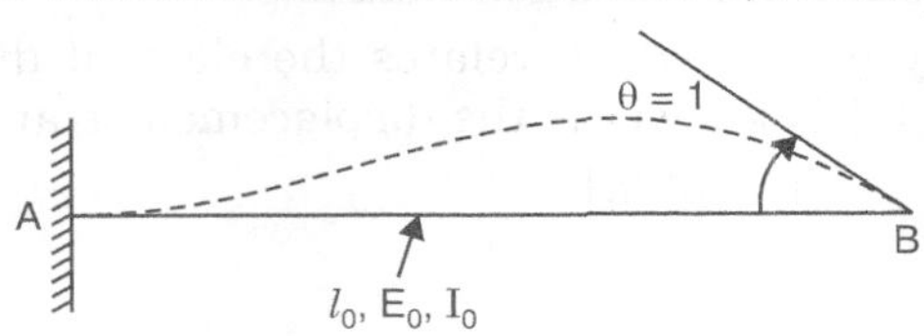

Fig. 6

Ans. $M_A = \dfrac{2E_0I_0}{l_0}$ clockwise

$M_B = \dfrac{4E_0I_0}{l_0}$ clockwise

$V_B = \dfrac{6E_0I_0}{l_0^2}$ upwards

18. For a portal frame fixed at the supports, which of these two methods is best suited for analysis-flexibility method or stiffness method ?

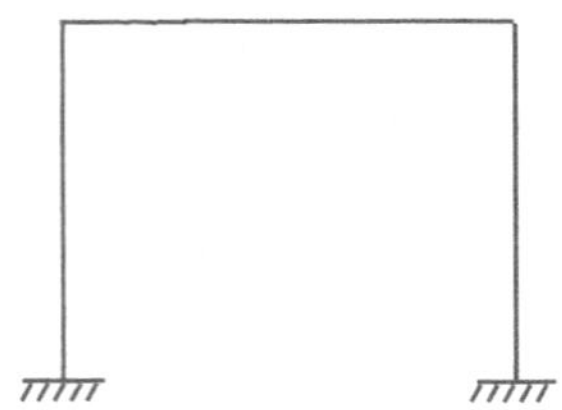

Fig. 7

Ans. Static indeterminacy = 3

Kinematic indeterminacy (including sway) = 3

Hence, both methods are equally suited.

19. Give the nodal degrees of freedom of pin jointed members of trusses.

Ans.

S.No.	*Structure type*	*D.O.F. per node*	*Remarks*
1.	Plane truss	2	Displacements along the 2 plane coordinate directions
2.	Space truss	3	Displacements along the 3 space coordinate directions.

20. Give the nodal degrees of freedom of rigid jointed frame members.

Ans.

S.No.	*Type of frame*	*D.O.F. per node*	*Remarks*
1.	Plane frame	3	Two displacements and one rotation
2.	Grid frame	3	One displacement and two rotations
3.	Space frame	6	Three displacements and three rotations

21. Define the displacement transformation matrix.

Ans. The connectivity matrix which relates the element displacement vector {δ} and system displacement vector $\{u\}$ is known as the displacement transformation matrix [β].

$$\{\delta\} = [\beta]\,\{u\}$$

22. Write down the stiffness matrix of the truss member shown in Fig. 8 (*a*).

Fig. 8 (a)

Ans.

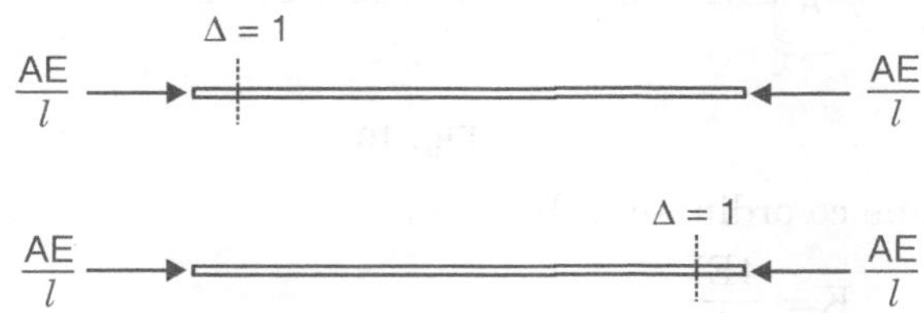

Fig. 8 (b)

In the question the co-ordinate direction for 2 is unconventional but admissible. Fig. 8 (*b*) shows the forces at 1 and 2.

(*i*) when we apply a unit displacement at 1 only (and arrest displacement at 2)

(*ii*) when we apply a unit displacement at 2 and arrest 1.

Hence, $$[K] = \frac{AE}{l}\begin{bmatrix} 1 & 1 \\ 1 & 1 \end{bmatrix}.$$

23. The [K] matrix in example 22 obviously has a determinant equal to zero. What is your inference about the element ?

Ans. If the determinant is zero, it shows the structure (in this case the element) is unstable. It is obvious that the element in Fig. 8 (*a*) is unstable and free to make rigid body displacements. It does not qualify to be called a **structure.**

24. What do you mean by global or system stiffness matrix ?

Ans. The stiffness matrix of a structure, obtained after transformation and assemblage of its element stiffness matrices along the global co-ordinate axes is termed as global or system stiffness matrix.

25. A propped cantilever beam has a uniform section, span l and flexural rigidity EI. What is the stiffness co-efficient corresponding to rotation of the propped end ?

Ans.

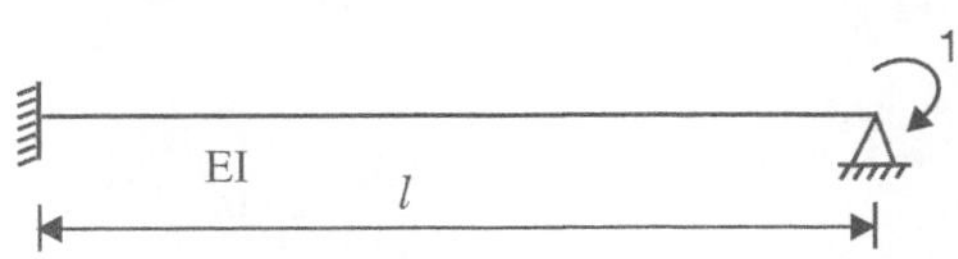

Fig. 9

1 is the co-ordinate corresponding to rotation of the propped end. The moment required to cause unit rotation along the coordinate, $k_{11} = \frac{4EI}{l}$

The stiffness coefficient, $k_{11} = \frac{4EI}{l}$

26. A propped cantilever carries a *udl* of *w*/unit length all over. Use matrix stiffness method to determine the slope at the prop.

Ans.

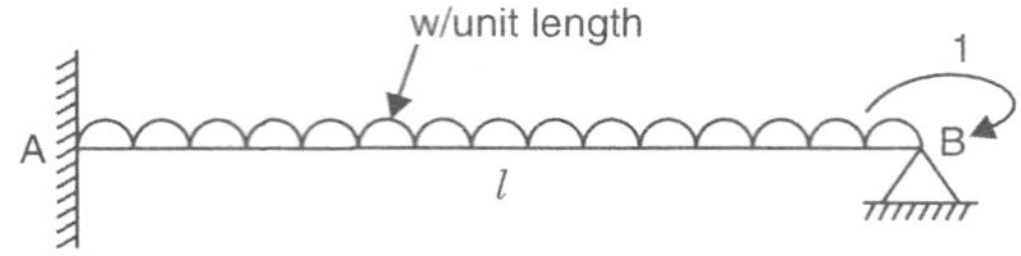

Fig. 10

Let us assign one co-ordinate at B. Then

$$K = \frac{4EI}{l}$$

The load w is not at the co-ordinate. The effect of w at 1 is

$$\{F\} = -\,[\text{FEM at B}] = -\frac{wl^2}{12}$$

$$\{F\} = [K]\,\{u\}$$

$$\{u\} = [K]^{-1}\,\{F\}$$

$$\{u\} = \frac{l}{4EI}\cdot\left(-\frac{wl^2}{12}\right) = -\frac{wl^2}{48EI}.$$

$\{u\}$ is the rotation at 1. Since $\{u\}$ is negative, the rotation is anti-clockwise.

27. Derive from first principles, the first column of the stiffness matrix of the 4 co-ordinate element shown in Fig. 11 (*a*).

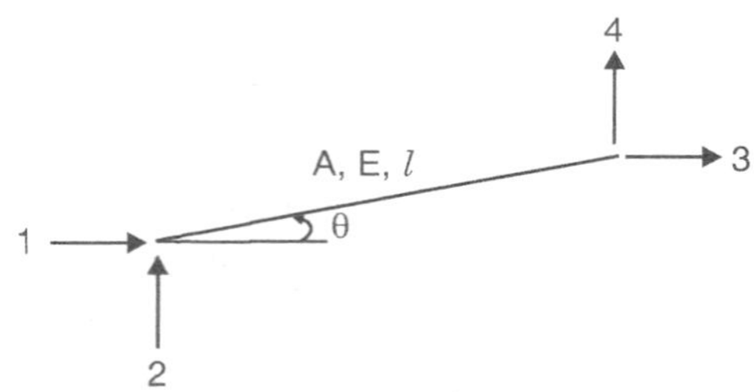

Fig. 11 (a)

Ans.

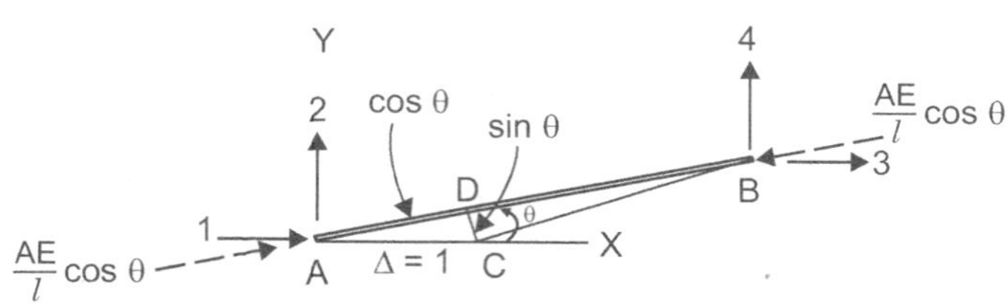

Fig. 11 (b)

AC = Δ = 1 → applied displacement at 1.

CB is the deformed element.

Deformation = AD = cos θ

The corresponding forces at A and B are (as shown) $\frac{AE}{l}\cos\theta$. Their components along 1, 2, 3 and 4 make the first column of [K] and are given by

$$\frac{AE}{l}\begin{bmatrix}\cos^2\theta & \cos\theta\sin\theta & -\cos^2\theta & -\cos\theta\sin\theta\end{bmatrix}^T$$

28. Why is a stiffness matrix a square matrix ?

Ans. If there are n co-ordinates, any column i contains n forces at the n co-ordinates when a unit displacement is applied at i (alone). There are n such co-ordinates at which unit forces have to be applied. Hence, there are n columns in a [K] matrix, each containing n elements. So [K] is of size $n \times n$.

29. While all flexibility matrices have their inverse, viz. the stiffness matrices, all stiffness matrices need not have their inverse, *i.e.*, flexibility matrices, why ?

Ans. We can generate stiffness matrices for deficient and unstable elements (and systems) also. This is possible because, when we generate any column of a [K] matrix, we assume passive forces at all co-ordinates except one.

For a deficient system, flexibility matrix cannot be generated. This is because, to develop a $[a]$ matrix we must apply a force at one co-ordinate only. This, in a deficient structure will cause infinite displacement(s). So $[a]$ can not exist.

The [K] matrix exists for an unstable system; but it cannot be inverted to get the $[a]$ matrix. The [K] matrix will be 'singular'.

30. What is the key equilibrium equation in analysing structures by matrix stiffness method ?

Ans. The matrix equation

$\{F\} = [K]\{u\}$ is called the equilibrium equation. Herein,

{F} is the matrix of system forces

[K] is the system stiffness matrix

and $\{u\}$ the matrix of system displacements

$\{u\}$ would normally include zero displacements at nodes restrained.

31. The stiffness matrix of a beam is given as $\begin{bmatrix}4 & 2\\ 2 & 4\end{bmatrix}$. When the nodal forces are $\begin{Bmatrix}10\\ 8\end{Bmatrix}$, find the nodal displacements.

Ans.

$$[k]\{u\} = [f]$$

$$\begin{bmatrix}4 & 2\\ 2 & 4\end{bmatrix}\begin{Bmatrix}u_1\\ u_2\end{Bmatrix} = \begin{Bmatrix}10\\ 8\end{Bmatrix}$$

$$\begin{Bmatrix}u_1\\ u_2\end{Bmatrix} = \begin{bmatrix}4 & 2\\ 2 & 4\end{bmatrix}^{-1}\begin{Bmatrix}10\\ 8\end{Bmatrix} = \frac{1}{(16-4)}\begin{bmatrix}4 & -2\\ -2 & 4\end{bmatrix}\begin{Bmatrix}10\\ 8\end{Bmatrix}$$

$$\begin{Bmatrix}u_1\\ u_2\end{Bmatrix} = \frac{1}{12}\begin{bmatrix}24\\ 12\end{bmatrix} = \begin{Bmatrix}2\\ 1\end{Bmatrix}$$

32. $[K_1]$ and $[K_2]$, the stiffness matrices of elements 1 and 2 are given below. Assemble them to get the stiffness matrix of the 3 co-ordinate system shown.

$$K_1 = \begin{bmatrix} 2 & -1.5 \\ -1.5 & 2 \end{bmatrix} \quad K_2 = \begin{bmatrix} 1.5 & -1 \\ -1 & 1.5 \end{bmatrix}$$

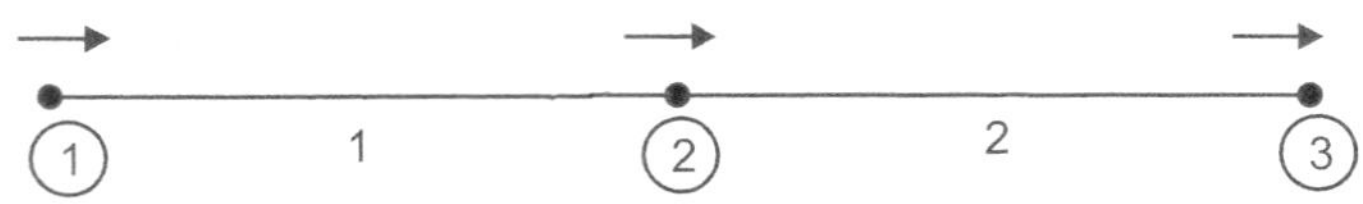

Fig. 12

Ans.

[K] =

2	– 1.5	
– 1.5	2 +1.5	– 1
	– 1	1.5

$$= \begin{bmatrix} 2 & -1.5 & 0 \\ -1.5 & 3.5 & -1 \\ 0 & -1 & 1.5 \end{bmatrix}$$

33. The stiffness matrices of elements 1 and 2 shown in Fig. 13 above are given by

$$[K_1] = \begin{bmatrix} 12 & 6 \\ 6 & 12 \end{bmatrix} \qquad [K_2] = \begin{bmatrix} 16 & 8 \\ 8 & 16 \end{bmatrix}$$

Assemble them to get the [K] for the beam ABC.

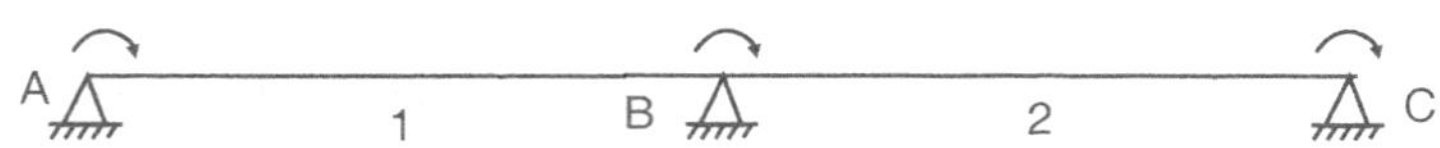

Fig. 13

Ans.

[K] =

12	6	
6	12 + 16	8
	8	16

$$= \begin{bmatrix} 12 & 6 & 0 \\ 6 & 28 & 8 \\ 0 & 8 & 16 \end{bmatrix}$$

34. For the beam shown in Fig. 14 (*a*), derive column 1 of the stiffness matrix.

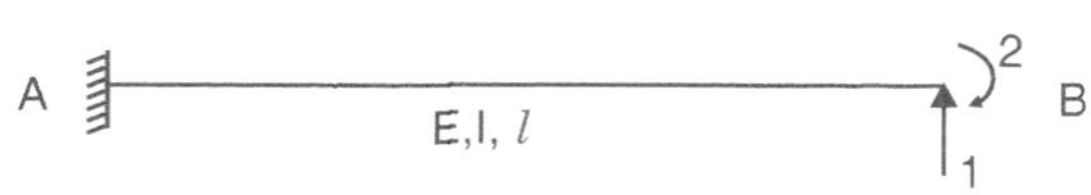

Fig. 14 (a)

Ans.

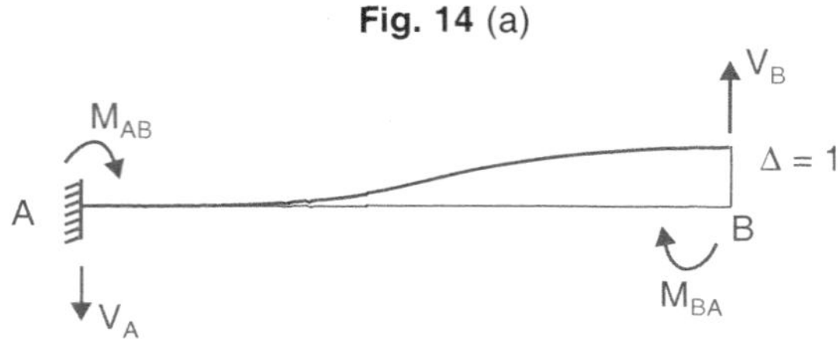

Fig. 14 (b)

When we apply $\Delta = 1$ at B,

$$M_{AB} = M_{BA} = \frac{6EI}{l^2}$$

From equilibrium conditions

$$V_B l = \frac{12EI}{l^2}; \qquad V_B = \frac{12EI}{l^3}$$

Hence the first column of $[k]$ is $\begin{Bmatrix} V_B \\ M_{AB} \end{Bmatrix} = \begin{Bmatrix} 12EI/l^3 \\ 6EI/l^2 \end{Bmatrix}$.

35. For the beam shown in Fig. 14 (*a*), derive the second column of $[k]$.

Ans.

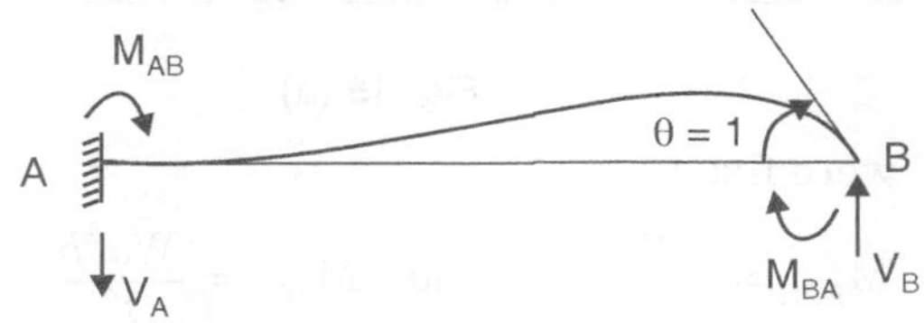

Fig. 14 (c)

When we apply $\theta = 1$ at B [Fig. 14 (*c*)]

$$M_{BA} = \frac{4EI}{l}; \quad M_{AB} = \frac{2EI}{l}$$

From equilibrium conditions, $V_B = V_A = \dfrac{6EI}{l^2}$

Hence column 2 of $[K] = \begin{bmatrix} 6EI/l^2 \\ 4EI/l \end{bmatrix}$.

36. A propped cantilever beam as shown in Fig. 14(*a*) has a stiffness matrix of

$$[K] = \begin{bmatrix} 120 & 60 \\ 60 & 40 \end{bmatrix}$$

Find the force at prop B when the prop sinks by 0.001 unit.

Ans.
$$F = [K]\{u\} = \begin{bmatrix} k_{11} & k_{12} \\ k_{21} & k_{22} \end{bmatrix}\{u\} = \begin{bmatrix} 120 & 60 \\ 60 & 40 \end{bmatrix}\{u\}$$

$$u = \begin{Bmatrix} u_1 \\ u_2 \end{Bmatrix} = \begin{Bmatrix} -0.001 \\ u_2 \end{Bmatrix}$$

F_1 is unknown ; F_2 is zero.

$$F_1 = 120u_1 + 60u_2 = -0.12 + 60u_2$$
$$F_2 = 60u_1 + 40u_2 = -0.06 + 40u_2 = 0$$
$$40u_2 = +0.06 \; ; \; u_2 = 0.0015$$
$$F_1 = -0.12 + 60 \times 0.0015 = 0.03 \text{ units.}$$

37. A fixed beam AB develops a rotation $\theta = 0.02$ radian and sinking by 0.01 units. Find the force components at B, assuming that the stiffness is as given in question 36.

Ans.
$$F = [k]\{u\} = \begin{bmatrix} 120 & 60 \\ 60 & 40 \end{bmatrix}\begin{Bmatrix} -0.01 \\ 0.02 \end{Bmatrix}$$

$$\begin{bmatrix} F_1 \\ F_2 \end{bmatrix} = \begin{Bmatrix} 1.2 + 1.2 \\ 0.6 + 0.8 \end{Bmatrix} = \begin{Bmatrix} 0 \\ 0.2 \end{Bmatrix}.$$

38. The stiffness matrix of an element BC of a frame (Fig. 15 *a*) is $[K] = \begin{bmatrix} 80 & 40 \\ 40 & 80 \end{bmatrix}$.

Get the equivalent vector of forces at 1 and 2 due to a load of 80 kN acting at D, 4 m from B.

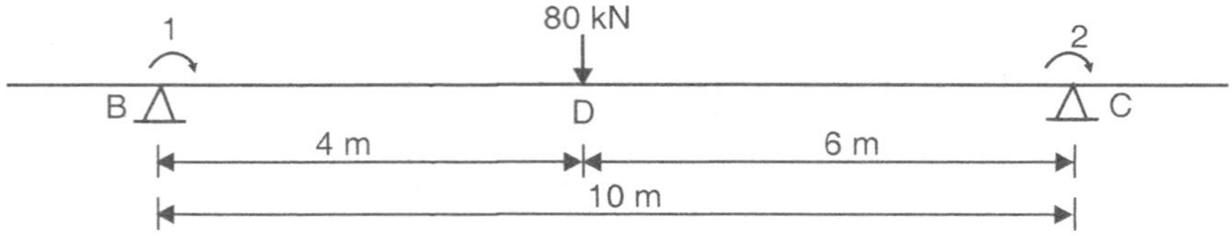

Fig. 15 (a)

Ans. If B and C were fixed,

$$M_{FBC} = -\frac{Wab^2}{l^2} \quad \text{and} \quad M_{FCB} = \frac{Wa^2b}{l^2}$$

$$M_{FBC} = -\frac{80 \times 4 \times 36}{100} = -115.2$$

$$M_{FCB} = \frac{80 \times 4^2 \times 6}{100} = 76.8$$

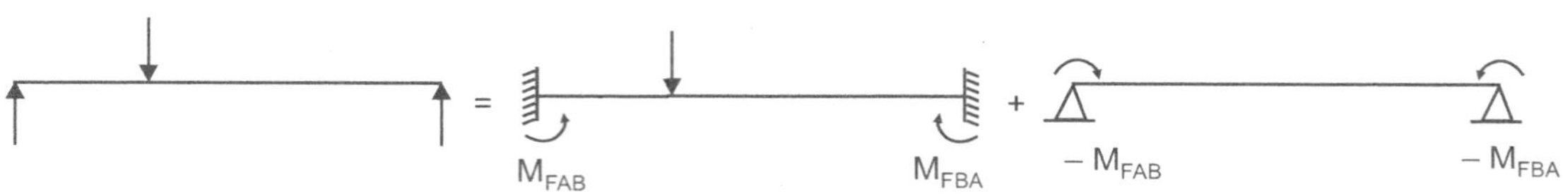

(b) (c) (d)

Fig. 15 (b), (c), (d)

During the matrix analysis, the force in Fig. 15 (*b*) will be replaced by the forces in Fig. 15 (*d*). Hence

$$\{F\}^e = \begin{Bmatrix} F_1 \\ F_2 \end{Bmatrix} = \begin{Bmatrix} 115.2 \\ -76.8 \end{Bmatrix}$$

where $\{F\}^e$ is the force vector equivalent to the given load of 80 kN at D.

Forces in Fig. 15 (*c*) would be added to the forces after the matrix analysis process.

39.

(a) System 1 (b) System 2

Fig. 16

Given that $[k_1] = \begin{bmatrix} 8 & 4 \\ 4 & 3 \end{bmatrix}$. Compute $[k_2]$.

Ans. $\begin{Bmatrix} F_1 \\ F_2 \end{Bmatrix} = [k_1] \begin{Bmatrix} u_1 \\ u_2 \end{Bmatrix}$ for the system (*a*).

Hence if $F_1 = 0$, this tends to become system 2.

$$\begin{Bmatrix} 0 \\ F_2 \end{Bmatrix} = \begin{bmatrix} k_{11} & k_{12} \\ k_{21} & k_{22} \end{bmatrix} \begin{Bmatrix} u_1 \\ u_2 \end{Bmatrix}$$

$$0 = k_{11} u_1 + k_{12} u_2$$

$$k_{11} u_1 = - k_{12} u_2$$

$$u_1 = - k_{11}^{-1} k_{12} u_2$$

In our case, $k_{11} = 8, \quad k_{11}^{-1} = \frac{1}{8}$

$$u_1 = -\frac{1}{8} \times 4 \times u_2 = -\frac{u_2}{2}$$

$\therefore$ $$F_2 = k_{21} u_1 + k_{22} u_2 = 4\left(-\frac{1}{2} u_2\right) + 3u_2$$

$$F_2 = [1] u_2$$

Hence, $[k_2] = [1]$ in system 2.

40. Referring to Fig. 17, if $\{\delta\} = [T]\{u\}$, express F in terms of P.

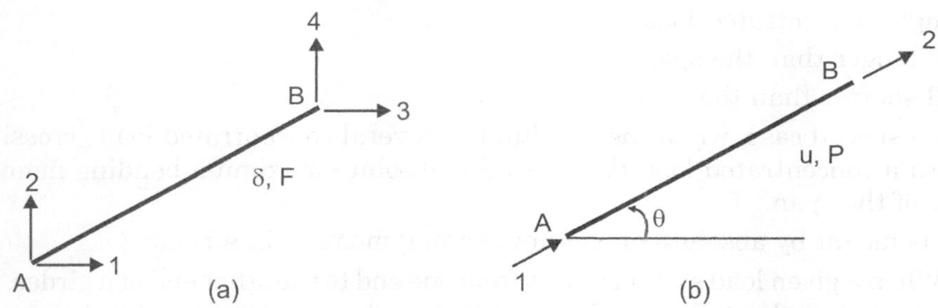

Fig. 17

Ans. $$\{\delta\} = [T]\{u\}$$

Since AB is the same member in systems 1 and 2, the work done on the system due to a set of forces P_1 and P_2 in system (*b*) is the same in system (*a*) also.

$$\delta^T F = u^T P$$

Since $\delta = [T]\{u\}, \quad \{\delta\}^T = u^T T^T$

So, $$u^T T^T F = u^T P$$

$$T^T F = P.$$

9

Rolling Loads

1. Where do you get rolling loads in practice ?

Ans. Shifting of load positions is common enough in buildings. But they are more pronounced in bridges and in gantry girders over which vehicles keep rolling.

2. Name the type of rolling loads for which the absolute maximum bending moment occurs at the midspan of a beam.

Ans. Rolling loads for which the absolute maximum bending moment occurs at midspan are:

(*i*) Single concentrated load

(*ii*) udl longer than the span

(*iii*) udl shorter than the span

(*iv*) As a special case, when the resultant of several concentrated loads crossing a span, coincides with a concentrated load then also the absolute maximum bending moment occurs at the centre of the span.

3. What is meant by absolute maximum bending moment in a beam ?

Ans. When a given load system moves from one end to the other end of a girder, depending upon the position of the load, there will be a maximum bending moment for every section. The maximum of these maximum bending moments will usually occur near or at the midspan. This maximum of maximum bending moments is called the absolute maximum bending moment, $M_{max\ max}$.

4. What is the absolute maximum bending moment that will occur in a simply supported beam when a udl shorter than the span moves from one end to the other end of it ?

Ans. Load position for $M_{max\ max}$ is shown in Fig. 1

$M_{max\ max}$ at midspan

$$= \frac{wc}{2} \times \frac{l}{2} - w \cdot \frac{c}{2} \cdot \frac{c}{4}$$

$$= \frac{wc}{4}\left(l - \frac{c}{2}\right)$$

$$M_{max\ max} = \frac{wc}{4}\left(l - \frac{c}{2}\right).$$

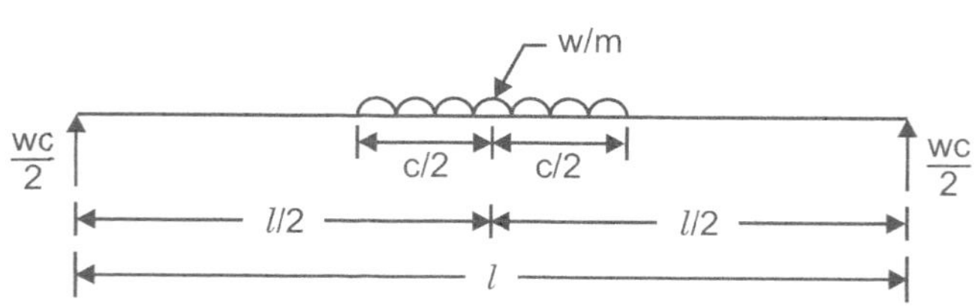

Fig. 1

5. When a udl shorter than the span passes over a simply supported beam, what is the load condition to get maximum bending moment at a point ?

Ans. Let c is the length of the udl.

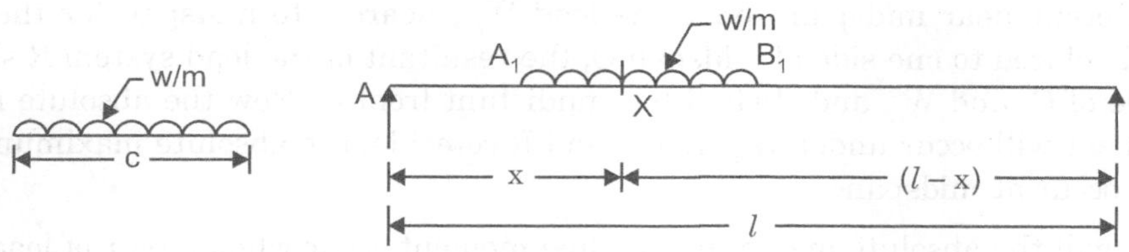

Fig. 2

For maximum bending moment at a given section X, the load should be placed in such a position that the average loading to the left of the section should be the same as the average loading to the right of the section.

$$\frac{A_1X}{XB_1} = \frac{AX}{XB} \quad i.e. \quad \frac{A_1X}{c - A_1X} = \frac{x}{(l - x)}$$

6. State the condition for maximum bending moment to occur at a given section of a simply supported beam due to a set of point loads moving on the span.

Ans. If we call the given section as X, in a span AB, then X divides the span in the ratio AX / XB. Normally there will be a certain load (the critical load) that shall be on the section X to cause maximum possible bending moment at X. The critical load is that which will change the rate of loading on AX and XB as shown in Fig 3.

Rate of loading on AX = $\frac{W_L}{x}$

Rate of loading on XB = $\frac{W_R}{(l - x)}$

Then let us consider the sign of R = $\left(\frac{W_L}{x} - \frac{W_R}{(l - x)}\right)$

If the critical load is to the left of X, R would be positive.

If the critical load is to the right of X, R would be negative.

This is the criterion to decide which is the critical load W_{cr}.

If W_{cr} is on X, maximum bending moment would occur at X.

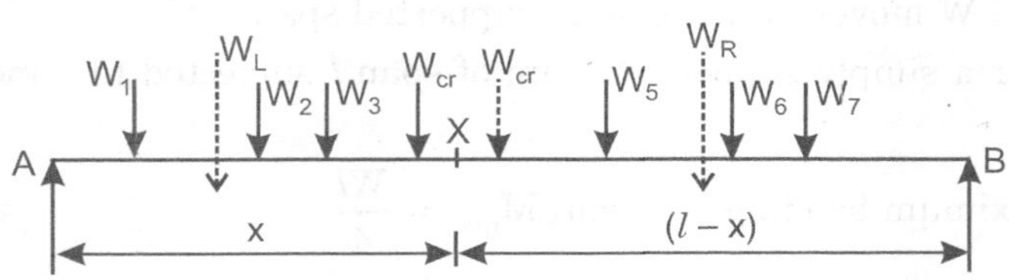

Fig. 3. Multiple point loads.

(The suspected W_{cr} shall be tried first left of X, then right of X. If R changes sign due the shift, then the W_{cr} chosen is correct.)

7. Where do you have the absolute maximum bending moment in a simply supported beam when a series of wheel loads cross it ?

Ans. When a series of wheel loads crosses a SS beam, the absolute maximum bending moment will occur near midspan under the load W_{cr}, nearest to midspan (or the heaviest load). If W_{cr} is placed to one side of midspan C, the resultant of the load system R shall be on the other side of C; and W_{cr} and R shall be equidistant from C. Now the absolute maximum bending moment will occur under W_{cr}. If W_{cr} and R coincide, the absolute maximum bending moment will occur at midspan.

8. Where will the absolute maximum bending moment occur when a pair of loads 100 kN and 60 kN spaced 2.4 apart cross a 16 m span.

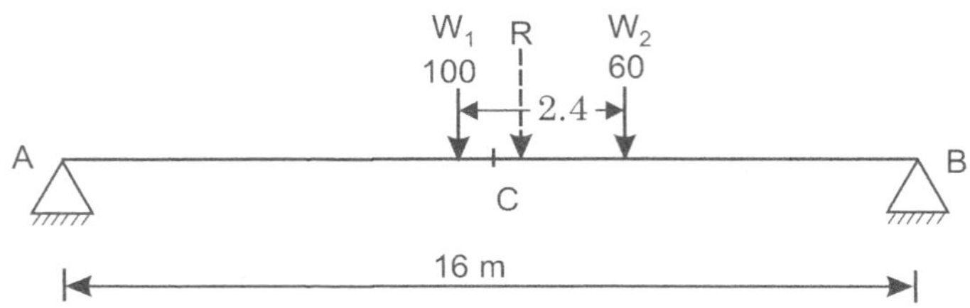

Fig. 4. Two rolling loads.

Ans. Resultant R will be $\frac{2.4 \times 60}{160} = 0.9$ m from W_1

The load position will be such that 100 kN load and R are equidistant from C, the mid-span.

100 kN load will be 8 – 0.45 m from A, *i.e.*, 7.55 m from A.

Maximum bending moment will occur at 7.55 m from A.

9. In the context of rolling loads, what do you understand by the term equivalent uniformly distributed load ?

Ans. A given system of rolling loads crossing a girder can always be replaced by a static uniformly distributed load of intensity w_e, such that the bending moment at any section under this static load is not less than that caused by the system of rolling loads. Such a static load is known as Equivalent Uniformly Distributed Load (EUDL).

10. What is the equivalent udl to cause the 'same bending moment' as when a single concentrated load W moves over a simply supported span?

Ans. Consider a simply supported beam of span l subjected to a moving concentrated load W.

Absolute maximum bending moment, $M_{max} = \frac{Wl}{4}$

If w_e is the equivalent udl, maximum bending moment $= \frac{w_e l^2}{8}$

Equating, $$\frac{w_e l^2}{8} = \frac{wl}{4}$$

$$w_e = \frac{wl}{4} \times \frac{8}{l^2} = \frac{2W}{l}$$

$$\text{EUDL, } w_e = \frac{2W}{l}$$

11. What is the practical use of determining equivalent udl ?

Ans. Finding the equivalent udl will help the designer. The designer can design the beams/slabs for a **static** equivalent udl without bothering to determine the critical positions of the moving live loads.

12. What is the absolute maximum bending moment due to a moving udl w longer than the span of a simply supported beam?

Ans. When a simply supported beam is subjected to a moving udl longer than the span, the absolute maximum bending moment occurs when the whole span is loaded.

$$M_{max} = \frac{wl^2}{8}$$

13. A SS beam has maximum bending moment ordinates as shown in Fig. 5. Determine the eudl.

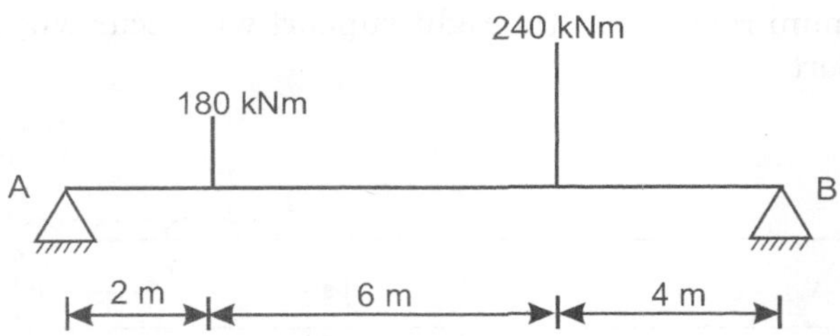

Fig. 5. Max BM ordinates.

Ans. Let us consider a parabola whose central ordinate is M_0. Span l = 12 m.

Equation to BMD is then $$M = \frac{4M_0}{l_0} x(l - x)$$

$$M = \frac{4}{144} x(12 - x) M_0$$

$$M_2 = 180 = \frac{1}{36} \times 2 \times 10 \times M_0 = \frac{10}{18} M_0$$

$$M_4 = 240 = \frac{1}{36} \times 4 \times 8 \times M_0 = \frac{8}{9} M_0$$

$$M_0 = \frac{180 \times 18}{10} = 364 \text{ kNm}$$

$$M_0 = \frac{240 \times 9}{8} = 270 \text{ kNm}$$

We have to take the bigger of the 2 values of M_0.

$$w_e = \frac{8M_0}{l^2} = \frac{8 \times 364}{144} = 20.22 \text{ kN/m.}$$

The result is shown grapically in Fig. 5(*a*)

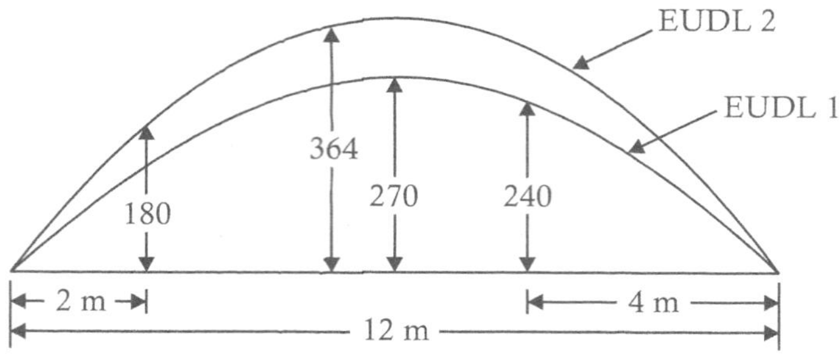

Fig. 5 (a) BMDs due to EUDLs.

14. State the location of maximum shear force in a simple beam with any kind of loading.

Ans. In a simple beam with any kind of load, the maximum positive shear force occurs at the left support and maximum negative shear force occurs at right support.

15. A udl of length shorter than the span crosses a girder. What should be the position of load to cause maximum reaction at the right support ?

Ans. When a udl shorter than the span crosses a simply supported a girder, say from left to right, the maximum reaction at the right support will occur when the head of the load reaches the right support.

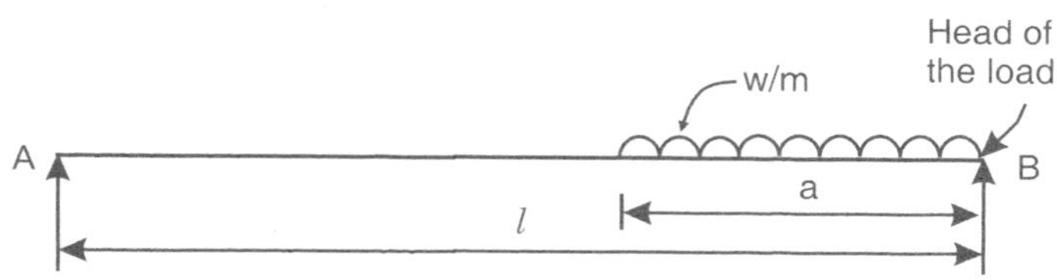

Fig. 6. Load position for maximum R_B.

16. Point P is 4 m from the left support of a simply supported beam of span 12 m. Locate a moving udl of length 3 m so as to produce the maximum bending moment at P.

Ans.

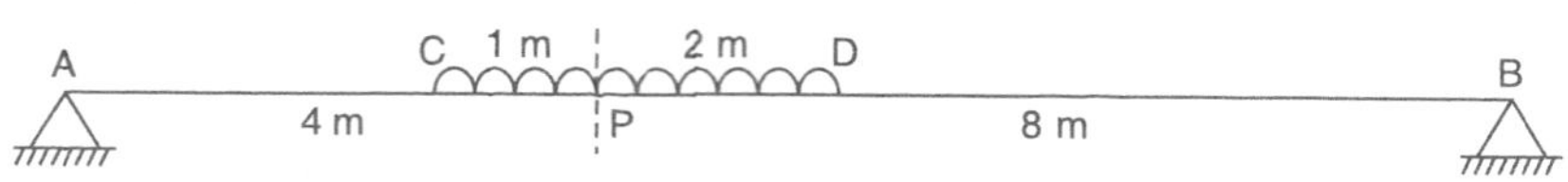

Fig. 7 (a)

Fig. 7(*a*) shows the critical load position. The rule is

$$AP : PB = CP : PD$$

Hence, CP = 1 m.

17. Define 'focal length' as applied to rolling loads.

Ans. In a SS beam, when the SFD due to dead load and the SFD due to a given system of live loads are combined there will be a portion of the beam on the left and on the right in which the net shear force is always positive or negative. There will be a middle portion in which the net shear force can be either positive or negative *i.e.,* as the load moves the shear force changes sign over a middle portion. This zone in a girder, in which the net shear force can change sign is called the 'focal length' [Fig. 7 (*b*)].

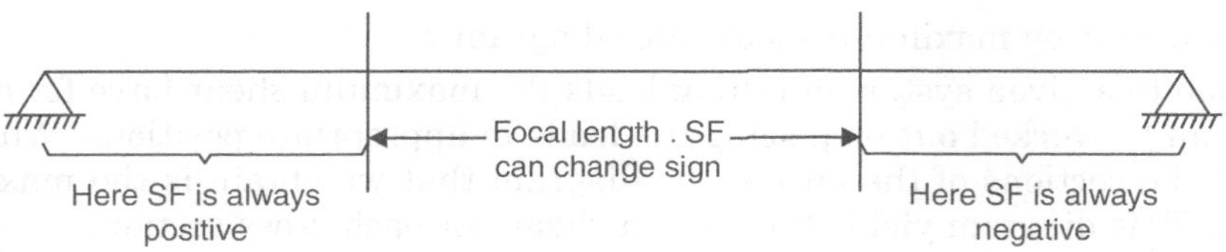

Fig. 7 (b)

18. A 9 m span 55 beam carries a uniform dead load of 4 kN/m. A concentrated load of 12 kN might traverse the span. Determine graphically the focal length of the span (region of change of shear).

Ans. Max DL shear = $9 \times \frac{4}{2} = 18$ kN.

Max LL shears = 12 kN (+ or –)

Fig. 7(*c*) shows the combinations and zone of shear change.

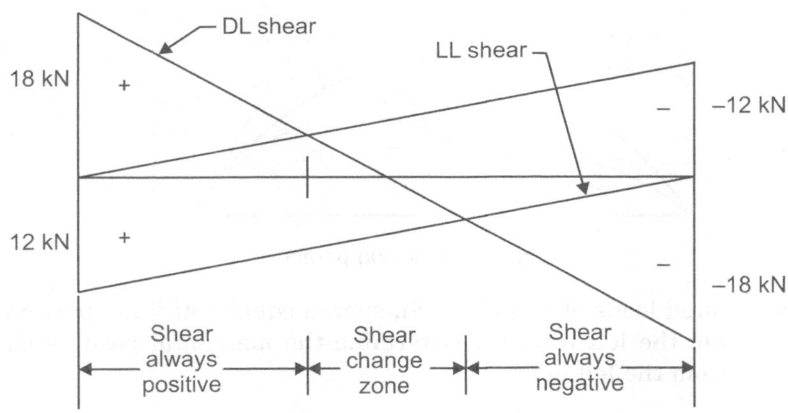

Fig. 7 (c) Shear change zone.

19. How will you place the load system to cause maximum bending moment under a given wheel load when a series of concentrated loads moves on a simply supported girder?

Ans. When a series of concentrated loads moves on a simply supported girder, the maximum bending moment under a given wheel load W_2 occurs when the given wheel load is so placed that the centre of the span is midway between the C.G of the load system and the wheel under which the bending moment is required (see Fig. 8).

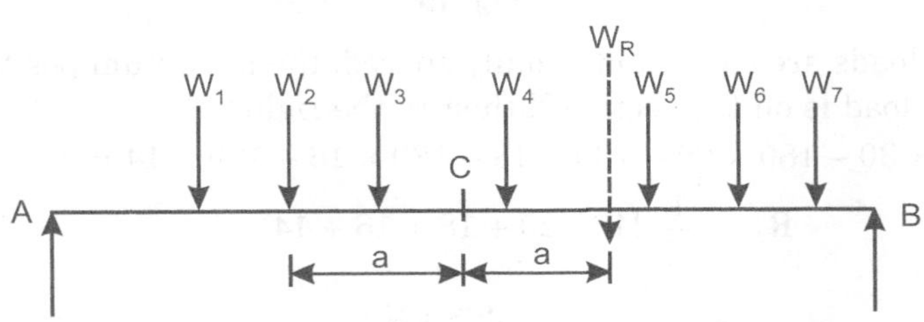

Fig. 8. Load position to cause maximum BM under W_2

20. What is meant by maximum shear force diagram ?

Ans. Due to a given system of rolling loads the maximum shear force for every section of the girder can be worked out by placing the loads in appropriate positions. When these are plotted for all the sections of the girder, the diagram that we obtain is the maximum shear force diagram. This diagram yields the 'design shear' for each cross-section.

21. What is meant by combined shear force diagram ?

Ans. When the SFD due to dead load and the SFD due to EUDL corresponding to a given system of loads, based on shear considerations are combined by superimposing them, the resulting diagram is called the combined shear force diagram.

22. What is meant by enveloping parabola ?

Ans. Enveloping parabola is usually drawn to determine the equivalent udl in the case of rolling loads on a SS girder. For example in Fig. 9, if the 2 parabolas A and B represent the maximum bending moment diagram due to 2 rolling loads, the enveloping parabola C will have ordinates at each and every point on the girder more than or equal to the maximum BMD ordinates.

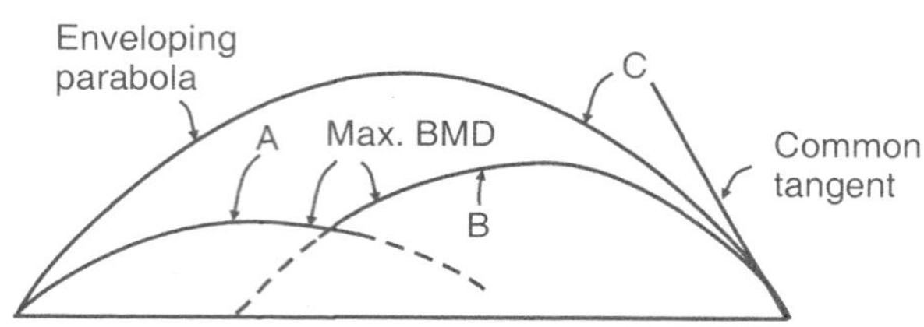

Fig. 9. Enveloping parabola.

23. Four concentrated loads of 180 kN each, spaced equally at 2 m apart cross a girder of span 30 m from the left to right. Determine the maximum positive shear force at a section 10 m from the left end.

Ans.

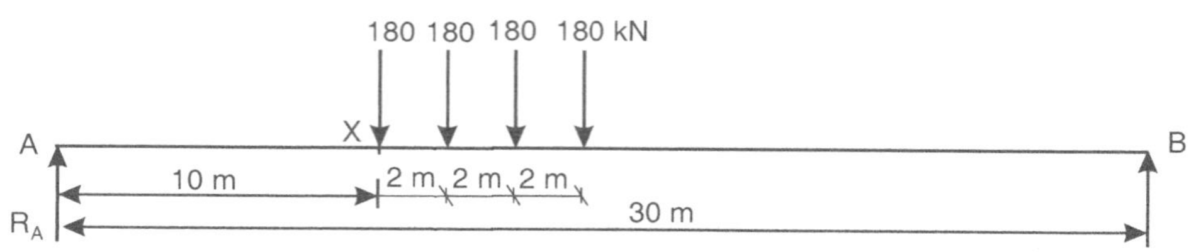

Fig. 10

Since all the loads are equal and equally spaced, the maximum positive shear at X will occur when the tail load is on the section X (just to the right).

$$R_A \times 30 - 180 \times 20 - 180 \times 18 - 180 \times 16 - 180 \times 14 = 0$$

$$R_A = \frac{1}{30} [180(20 + 18 + 16 + 14)]$$

$$= \frac{180}{30} \times 68 = 408 \text{ kN}$$

Max. positive shear force at X = + R_A = 408 kN

24. Position the 2 loads shown in Fig. 11 (*a*) to cause the maximum bending moment at support B in Fig. 11 (*b*).

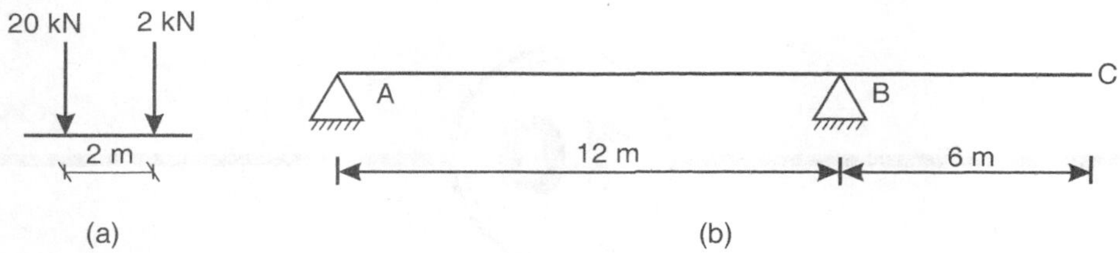

Fig. 11

Ans. The position for maximum bending moment at B will be one of the 2 positions drawn in Fig. 11 (*c*).

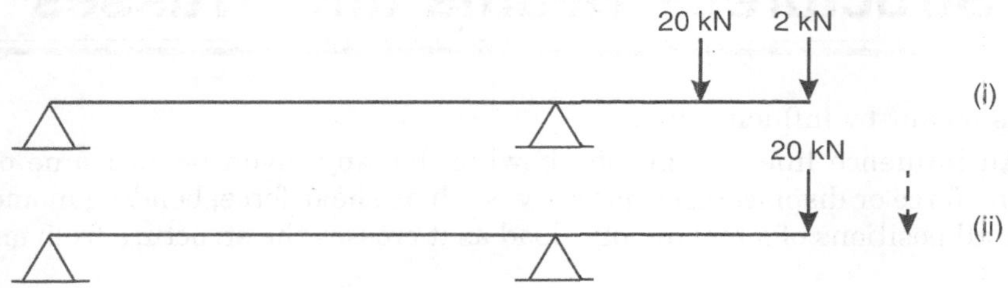

Fig. 11 (c)

Position (*i*) gives $M_B = 20 \times 4 + 2 \times 6 = 80 + 12 = 92$ kNm

Position (*ii*) gives $M_B = 20 \times 6 = 120$ kNm and is therefore the required load position.

10

Influence Lines for Statically Determinate Structures—Beams and Trusses

1. What is meant by influence lines ?

Ans. An influence line is a graph showing, for any given beam frame or truss, the variation of any force or displacement quantity (such as shear force, bending moment, tension, deflection) for all positions of a moving unit load as it crosses the structure from one end to the other.

2. Distinguish between the bending moment diagram for a beam and influence line diagram.

Ans. The ordinate of a BMD at any x gives the BM at that value of x, for any given loading.

ILD for BM is for a given point P in the beam. The ordinates of an ILD at any x give the bending moment at P due to a unit load at the point given by x.

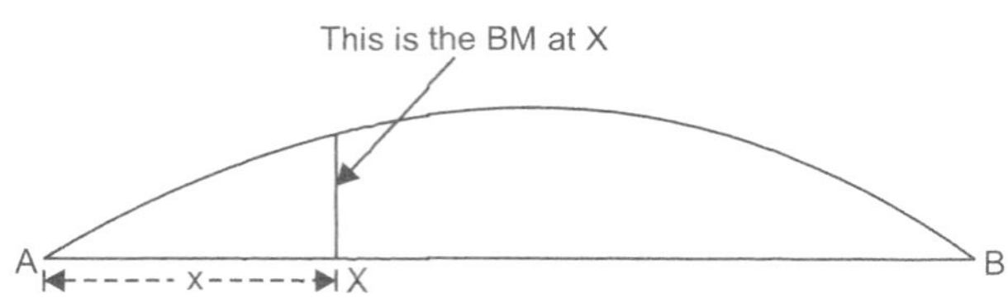

Fig. 1 (a). BMD

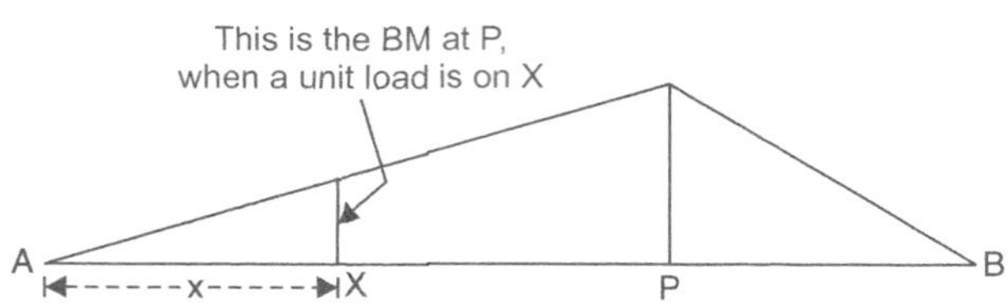

Fig. 1 (b). ILD (for BM at P)

3. Draw the influence line diagram for shear force at a point X in a simply supported beam AB of span 'l' m.

Ans.

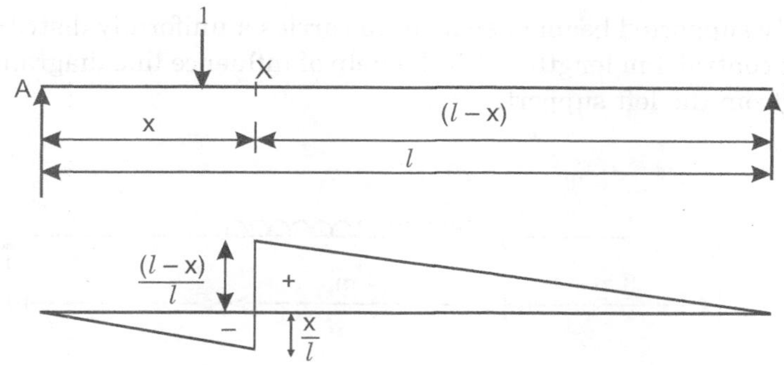

Fig. 1 (c) ILD for SF at a section X.

4. Draw the ILD for bending moment at any section X of a simply supported beam and mark the ordinates.

Ans.

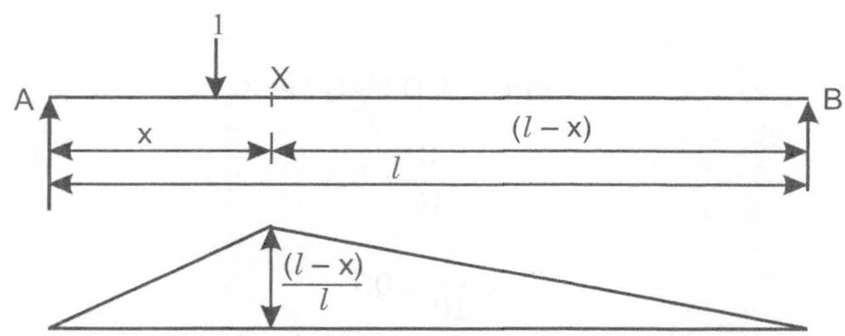

Fig. 2. ILD for BM at a section X.

5. What are the uses of influence line diagrams ?

Ans. (*i*) Influence lines are very useful in the quick determination of reactions, shear force, bending moment or similar functions at a given section under any given system of moving loads and

(*ii*) Influence lines are useful in determining the load position to cause maximum value of a given function in a structure on which load positions can vary.

6. Sketch the ILDs for the support reactions in a simply supported beam.

Ans.

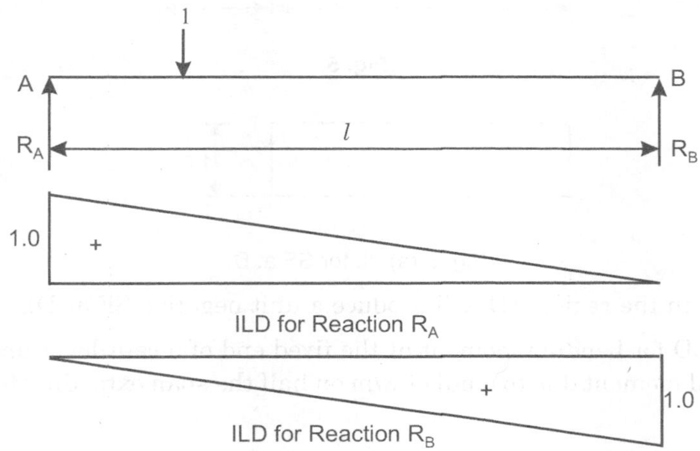

Fig. 3

7. A simply supported beam of span 10 m carries a uniformly distributed load of 20 kN/m over its central 4 m length. With the help of influence line diagram, find the shear force at 3 m from the left support.

Ans.

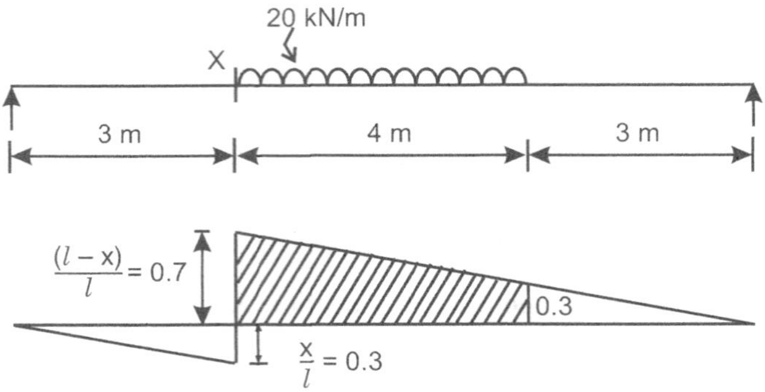

Fig. 4. ILD for SF at X.

$$\frac{x}{l} = \frac{3}{10} = 0.3$$

$$\frac{(l-x)}{l} = \frac{7}{10} = 0.7$$

Shear force at X = intensity of udl × area of ILD below the udl

$$= 20 \times \frac{(0.7 + 0.3)}{2} \times 4 = 40 \text{ kN}$$

8. Sketch the influence line for shear force at a section D in the cantilever shown in Fig. 5.

Fig. 5

Ans.

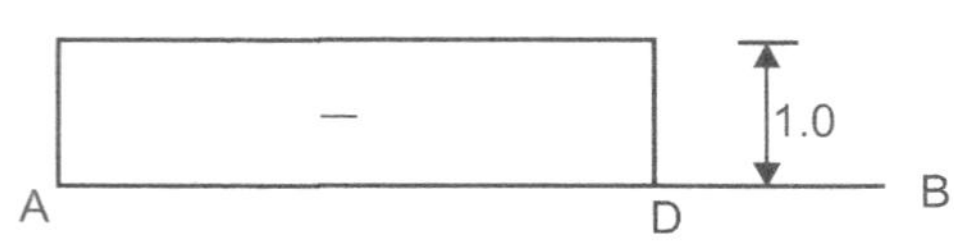

Fig. 5 (a) IL for SF at D.

(A unit load in the region AD will produce a unit negative SF at D).

9. Draw the ILD for bending moment at the fixed end of a cantilever and hence calculate the fixed end moment due to a udl of w/m on half the span extending from the fixed end.

Ans.

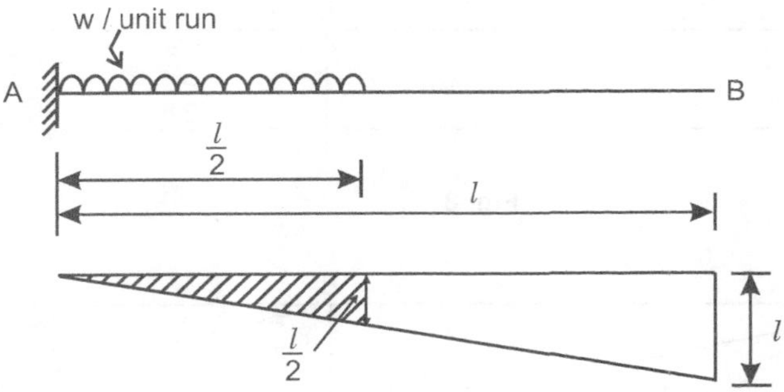

Fig. 6. ILD for BM at A.

$$\text{Fixed end moment} \quad = \frac{\left(0 + \frac{l}{2}\right)}{2} \times \frac{l}{2} \times w = \frac{wl^2}{8}$$

10. Sketch the influence line for reaction at the right support of a simply supported beam of span 8 m and hence determine the value of the maximum reaction when a udl of 6 kN/m and of length 5 m crosses the beam.

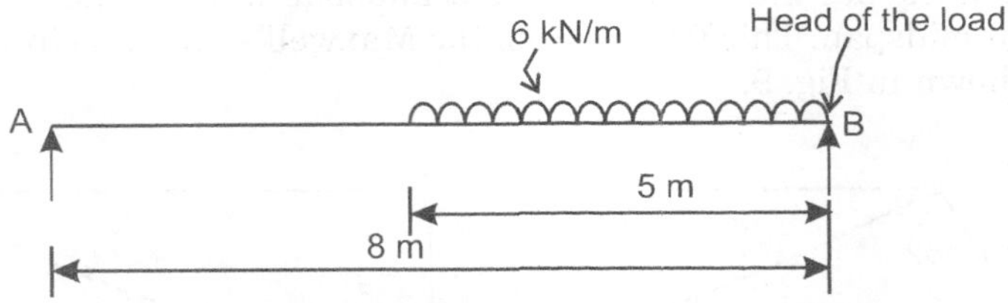

Beam with load position for max. R_B.

Ans.

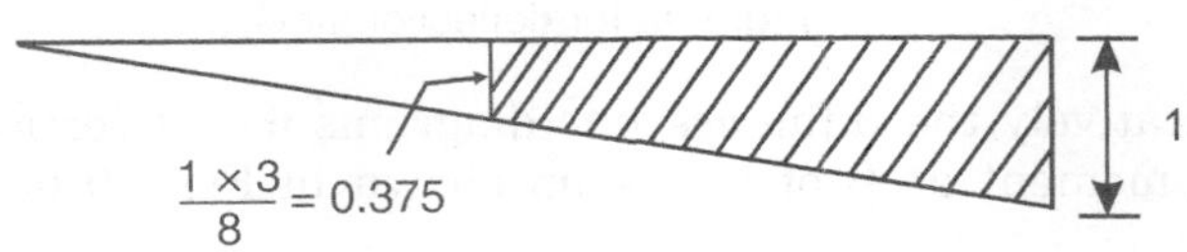

Fig. 7. ILD for R_B.

Max. R_B = Intensity of udl × area of ILD below udl

For Max. R_B, the head of the load is to be placed at B.

$$(R_B)_{max} = 6 \times \left(\frac{0.375 + 1}{2}\right) \times 5 = 20.625 \text{ kN}$$

11. For maximum bending moment at the fixed end of a cantilever, where will you place a single rolling load ?

Ans. A single rolling load shall be placed at the free end of the cantilever where the coordinate of ILD for bending moment at the fixed end will be the maximum.

12. For the overhanging beam shown in Fig. 8, sketch the influence lines for (*i*) shear to the left of B (F_{BL}), (*ii*) Shear to the right of B (F_{BR}).

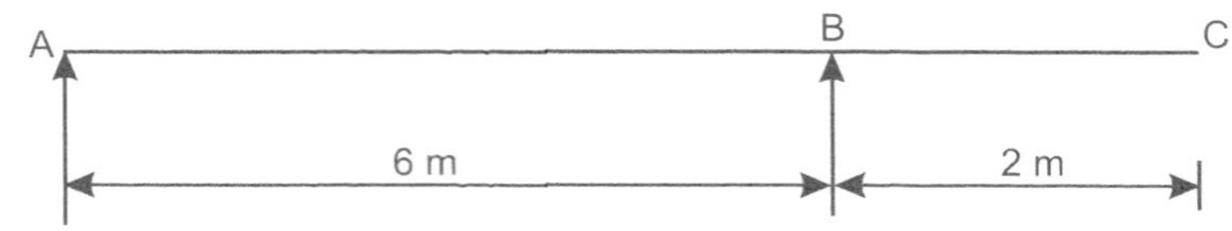

Fig. 8

Ans.

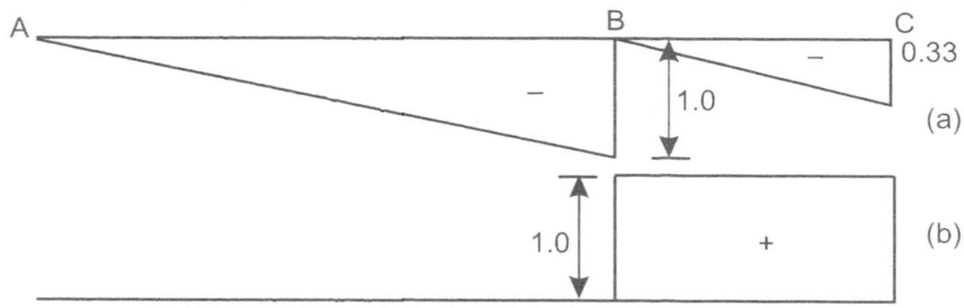

Fig. 8 (a) IL for F_{BL} ; (b) IL for F_{BR}.

13. Sketch the influence line for deflection at midspan in a simply supported beam of span l_0 and flexural rigidity E_0I_0.

Ans. The influence line for deflection at midspan is the same as the deflection diagram for a unit load at midspan. This follows from the Maxwell's law of reciprocal deflections. Hence, the ILD is as shown in Fig. 9.

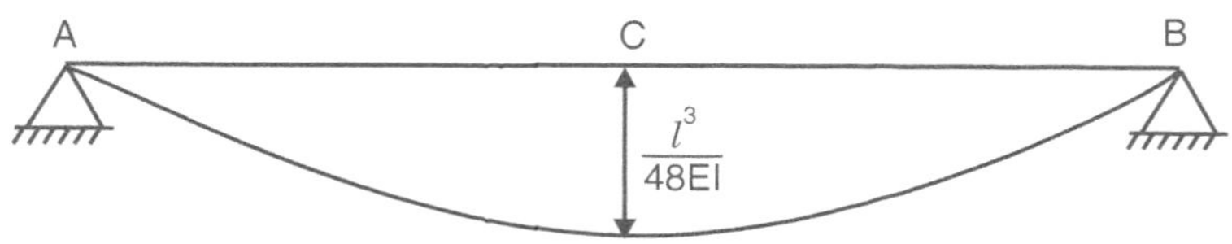

Fig. 9 IL for deflection at C.

14. Sketch qualitatively the influence line diagrams for (*i*) vertical reaction at support C (*ii*) Bending moment at B of the beam shown in Fig. 10 (*a*).

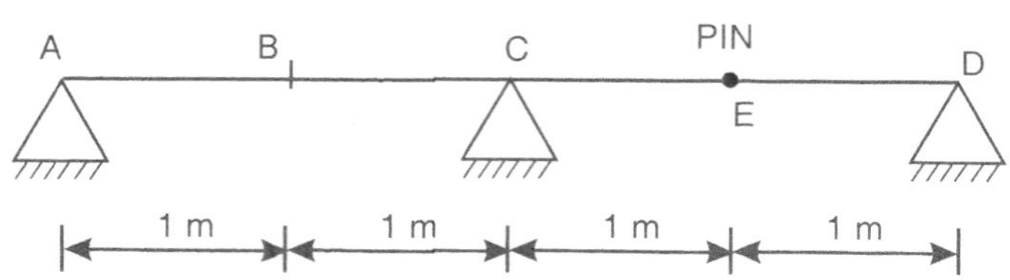

Fig. 10 (a)

Ans.

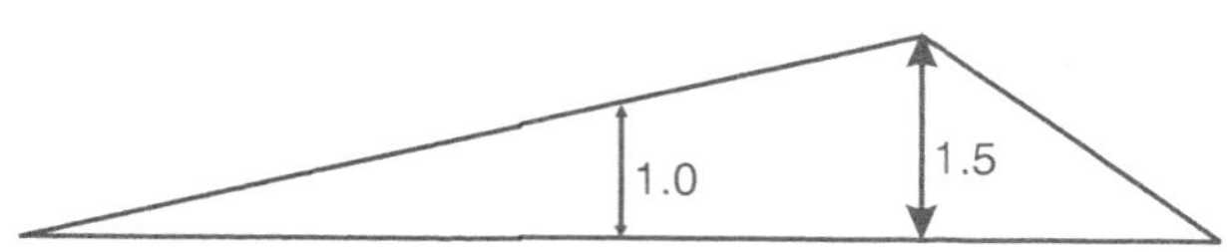

Fig. 10 (b) ILD for R_C.

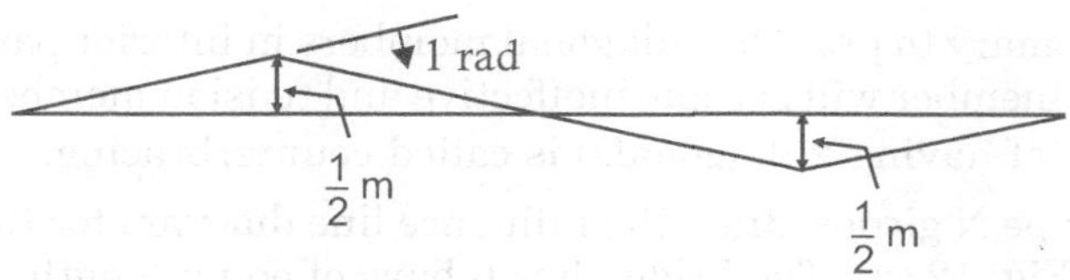

Fig. 10 (c) ILD for M_B.

15. Draw the ILD for reaction at B for the beam shown in Fig. 11 (*a*).

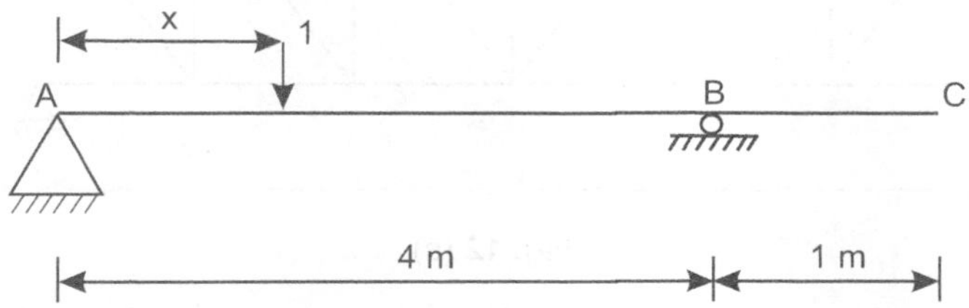

Fig. 11 (a)

Ans.

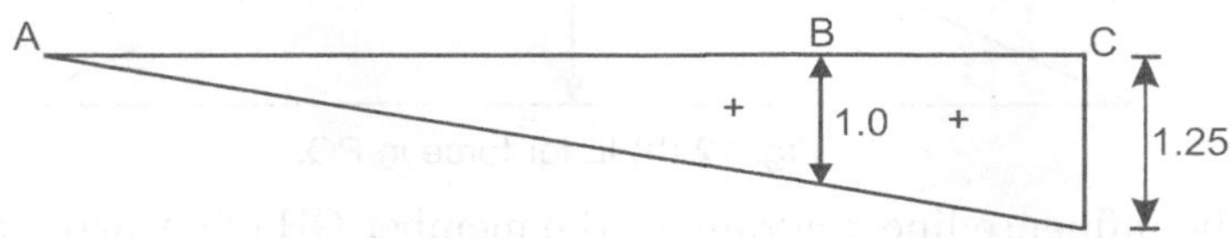

Fig. 11 (b) ILD for R_B.

Assume a moving unit load to be at a distance x from A. Taking moment about A,

$$R_B \times 4 - 1 \times x = 0$$

$$R_B = x/4$$

When $x = 0$, $R_B = 0$

When $x = 4$ m, $R_B = 4/4 = 1.0$

When $x = 5$ m, $R_B = 4/5 = 1.25$.

16. For the beam shown in Fig. 11 (*a*), what should be the position of a moving point load to cause maximum shear at B ?

Ans. Anywhere on BC.

17. What do you understand by the term 'reversal of stresses' ?

Ans. In certain long trusses the web members can develop either tension or compression depending upon the position of live loads. This tendency to change the nature of stresses is called reversal of stresses.

Members likely to get reversal of stresses shall be designed accordingly.

18. What is meant by counterbracing ?

Ans. In long trusses, web members near midspan can develop either tension or compression. Tension members are cheaper. Compression members have to be designed for buckling. They can be expensive.

Hence it is customary to provide 2 diagonal members in interior panels of bridge trusses. Here, the compressive member will become ineffective and tension member alone will be active.

This procedure (of having 2 diagonals) is called counterbracing.

19. For a through type N girder, draw the influence line diagram for the force in a top chord member PQ in Fig. 12 (*a*). The bridge has 6 bays of equal length.

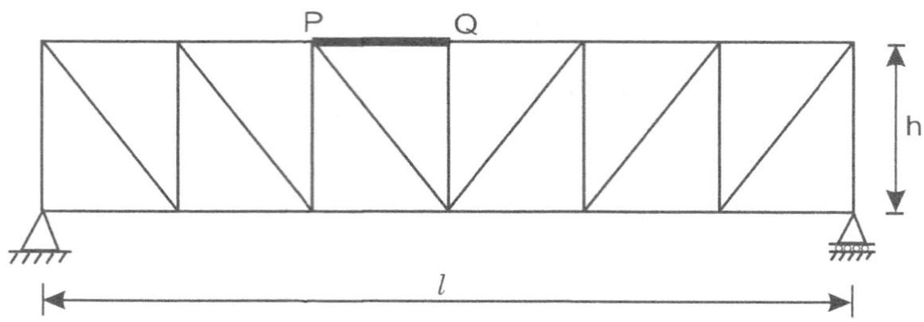

Fig. 12 (a)

Ans.

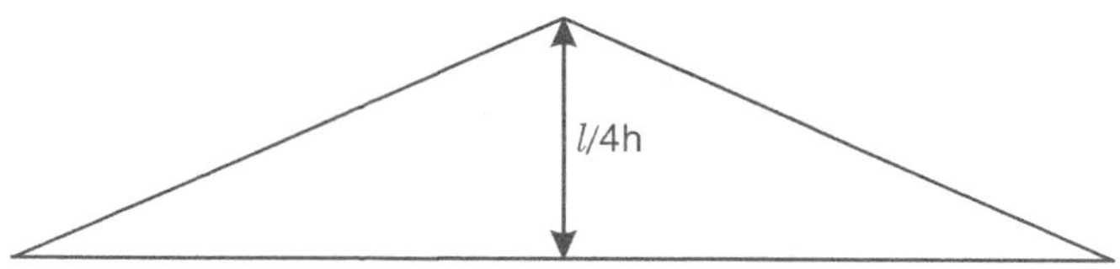

Fig. 12 (b) IL for force in PQ.

20. Sketch the influence line diagram for the member GH of a warren girder through truss shown in Fig. 13.

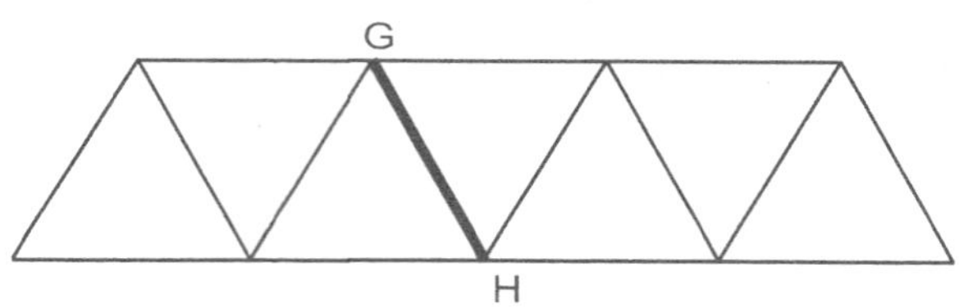

Fig. 13

Ans.

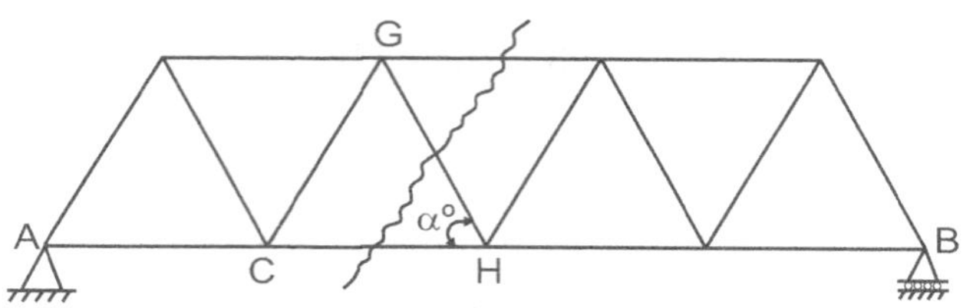

Fig. 13 (a)

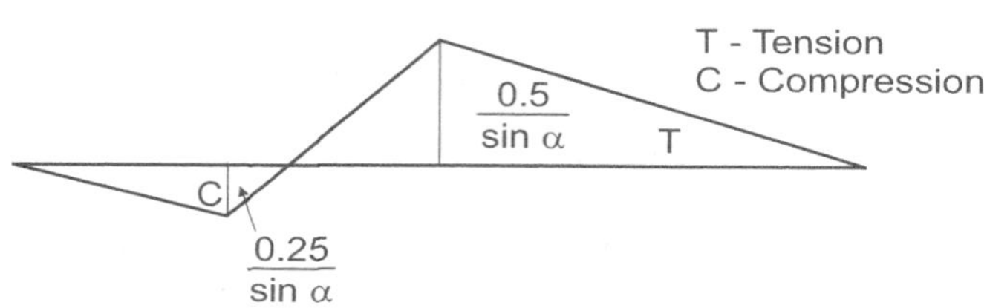

Fig. 13 (b) IL for force in GH.

(The vertical component of the force in GH shall take the shear force in the panel CH. Hence the force ordinates are got by dividing the shear force ordinate by sin α.)

21. In a parallel chord truss, which force parameter determines the force in a vertical member ?

Ans. Shear force in the panel.

22. Sketch the influence line diagram for the force in the member U_3L_3 of the through type girder shown in Fig. 14.

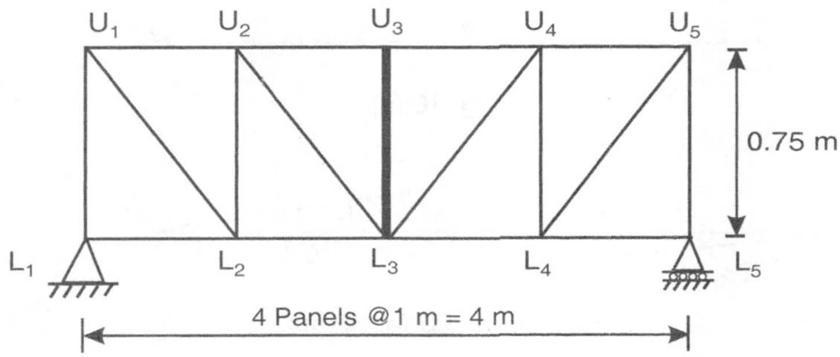

Fig. 14

Ans.

Fig. 14 (a)

Member U_3L_3 will have zero force, since this is a through bridge and there can be no force on the upper chord. Since, $U_2U_3U_4$ are collinear, U_3L_3 cannot have any force.

23. Sketch the influence line diagram for the force in member U_1L_1 of the through truss shown in Fig. 15.

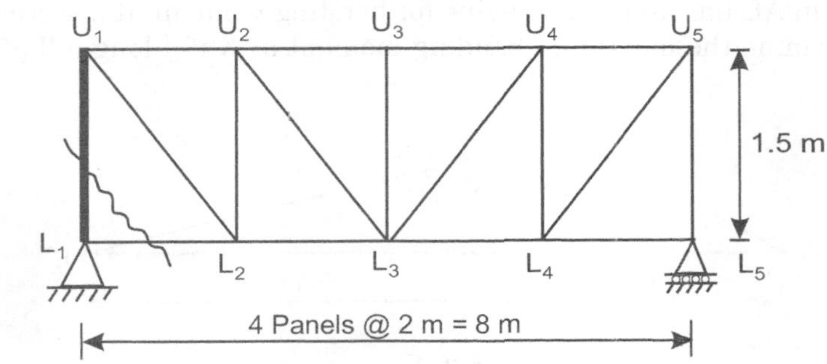

Ans.

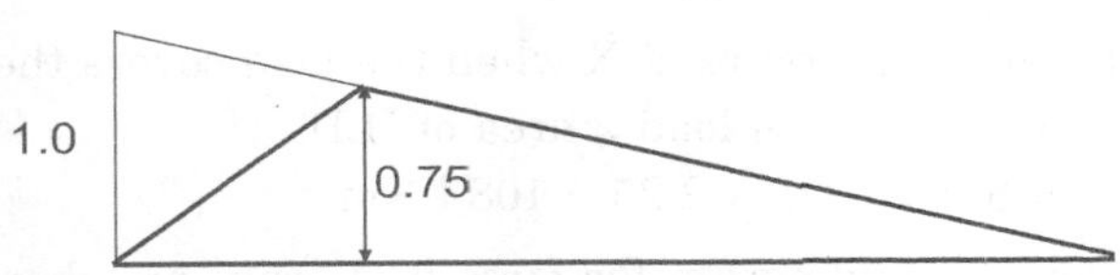

Fig. 15. IL for force in U_1L_1.
(Same as IL for shear in panel L_1L_2)

24. Draw the influence line diagram for shear force at C and hence determine shear force at C for the given loading. Refer to Fig. 16 (*a*).

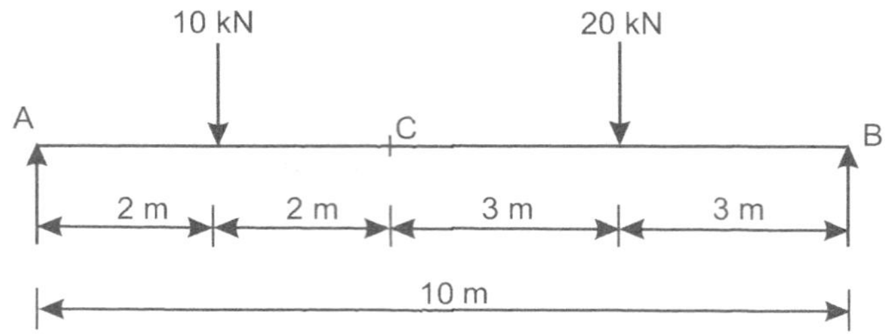

Fig. 16 (a)

Ans.

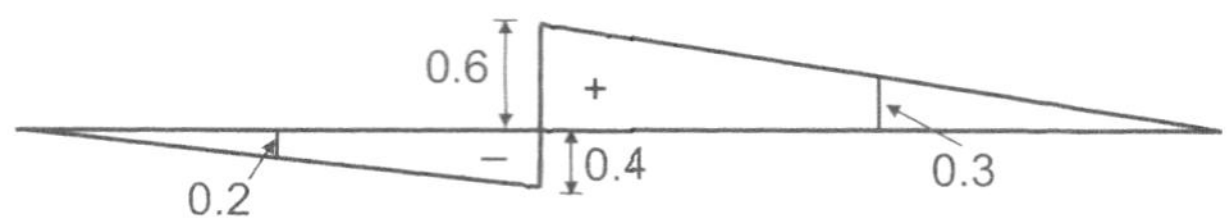

Fig. 16. (b) ILD for SF at C.

Max. +ve ordinate $= \frac{l - x}{l} = \frac{6}{10} = 0.6$

Max. –ve ordinate $= \frac{x}{l} = \frac{4}{10} = 0.4$

Ordinate under 10 kN load $= \frac{-0.4}{4} \times 2 = -0.2$

Ordinate under 20 kN load $= \frac{0.6}{6} \times 3 = 0.3$

Shear force at C $= 10 \times (-0.2) + 20 \times 0.3 = 4$ kN.

25. A certain beam AB has an influence line for bending moment at a section X as shown in Fig. 17. Determine the maximum bending moment at X if a long udl of 8 kN/m crosses the span.

Fig. 17

Ans. Maximum bending moment occurs at X when the load covers the entire span.

M_{max} = intensity of load × area of ILD

= 8 × ½ × 12 × 2.25 = 108 kNm.

26. Fig. 18 shows the influence line diagram for force in a truss member. Find the maximum tension in the member due to a dead load of 6 N/m and live load of 12 N/m, 12 m long.

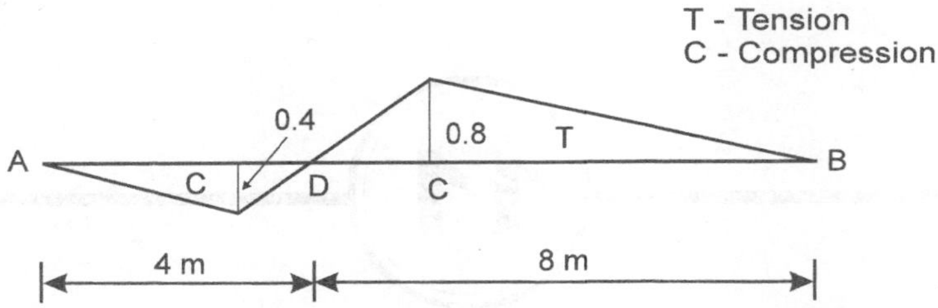

Fig. 18

Ans. For maximum tension in the member,

(*i*) the dead load will (of course) occupy the entire span AB.

(*ii*) The live load will occupy the portion DB.

$$T_{max} = 6\left(-\tfrac{1}{2} \times 0.4 \times 4 + \tfrac{1}{2} \times 0.8 \times 8\right) + 12\left(\tfrac{1}{2} + 0.8 \times 8\right)$$
$$= 14.4 + 38.4 = 52.8 \text{ kN.}$$

11

Influence Lines for Indeterminate Structures

1. State Muller-Breslau principle.

Ans. Muller-Breslau principle states that, if we want to sketch the influence line for any force quantity (like thrust, shear, reaction, support moment or bending moment) in a structure,

(*i*) We remove from the structure the restraint to that force quantity and

(*ii*) We apply on the remaining structure a unit displacement corresponding to that force quantity.

The resulting displacements in the structure are the influence line ordinates sought.

2. State Maxwell-Betti's theorem.

Ans. In a linearly elastic structure in static equilibrium acted upon by either of two systems of external forces, the virtual work done by the first system of forces in undergoing the displacements caused by the second system of forces is equal to the virtual work done by the second system of forces in undergoing the displacements caused by the first system of forces.

Maxwell Betti's theorem helps us draw influence lines for structures. It can be considered as a precursor of Muller-Breslau principle.

3. Sketch the qualitative influence line diagram for the reaction at the left support of a two span continuous beam having two equal spans simply supported at ends, by applying Muller Breslau principle.

Ans. Remove the support at A (Fig. 1).

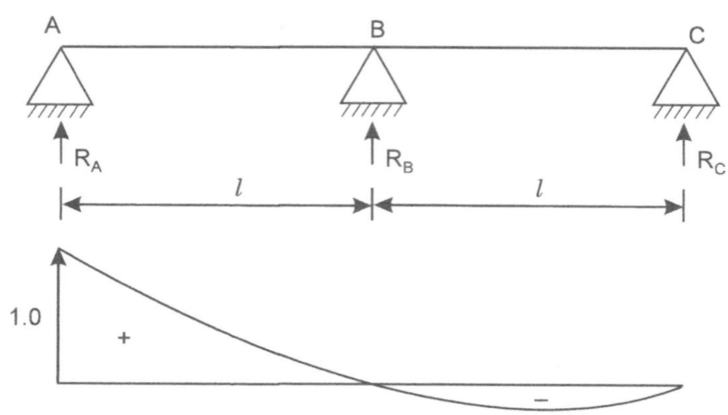

Fig. 1. ILD for R_A.

Apply a unit displacement to the beam at A in the direction of R_A. The resulting elastic curve is the influence line for R_A.

4. Sketch qualitatively the influence line diagram for shear force at any mid-span section of a two span continuous beam simply supported at all support points.

Ans. Remove only shear resistance of the beam at D. Give a cut at D and provide a sliding device so that the inclination of the beam will be the same on either side of D. (the resistance to bending moment of the beam at D will be intact).

Apply a unit shear displacement between the 2 parts of the beam on either side of the cut section. The resulting elastic curve of the beam is the influence line for SF at D (Fig. 2).

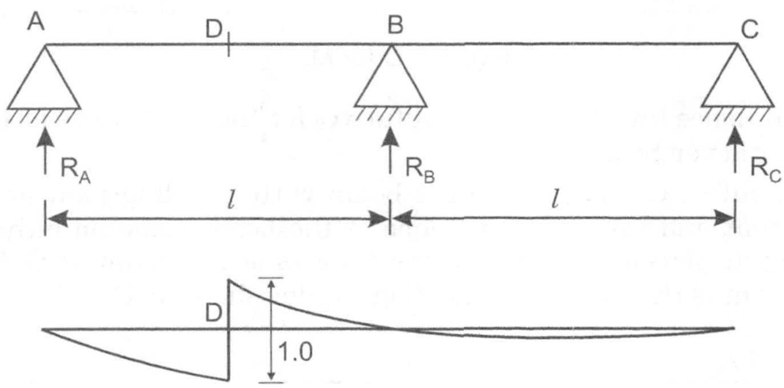

Fig. 2. ILD for SF_D.

5. Draw the influence line diagram (qualitative) for bending moment at the middle support of a continuous beam having two equal spans.

Ans. Remove the moment carrying capacity of the beam at B (by providing a hinge at B). Apply a unit rotation between the 2 parts of the beam on either side of B. The deflected line of the beam is the influence line diagram for M_B.

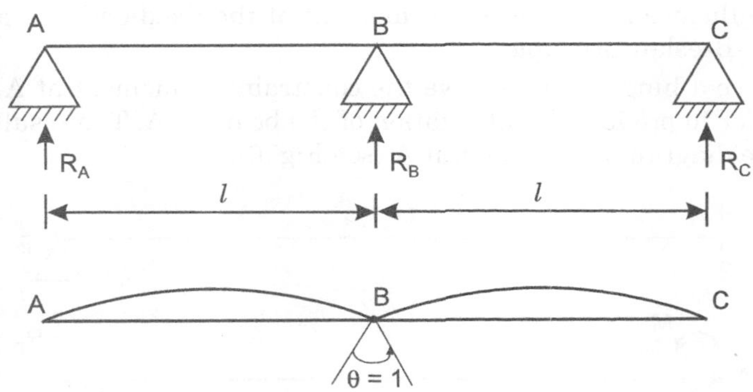

Fig. 3. ILD for M_B.

6. Sketch the influence line diagram (qualitative) for the bending moment at mid-span section of a propped cantilever beam.

Ans. Introduce a hinge at C (remove the restraint to bending moment). Apply at C, a pair of moments so as to produce a unit angular displacement between the 2 parts of the beam on either side of C. The resulting displaced line of the beam is the influence line for M_C.

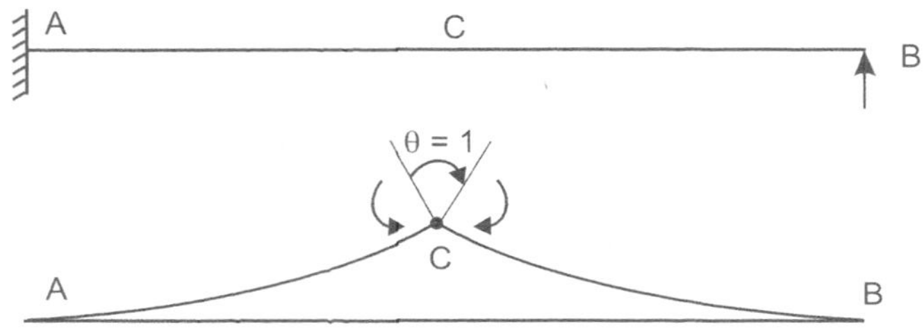

Fig. 4. ILD for M_C.

7. Sketch the influence line diagram (qualitative) for the shear force at midspan section of a propped cantilever beam.

Ans. Make a cut at C and provide the beam with a sliding joint at C. This joint will resist bending moment and maintain the slope of the beam same on either side of the cut. Apply a unit sliding displacement between the 2 parts of the beam at C. The resulting displaced line of the beam is the influence line diagram for shear at C (SF_C).

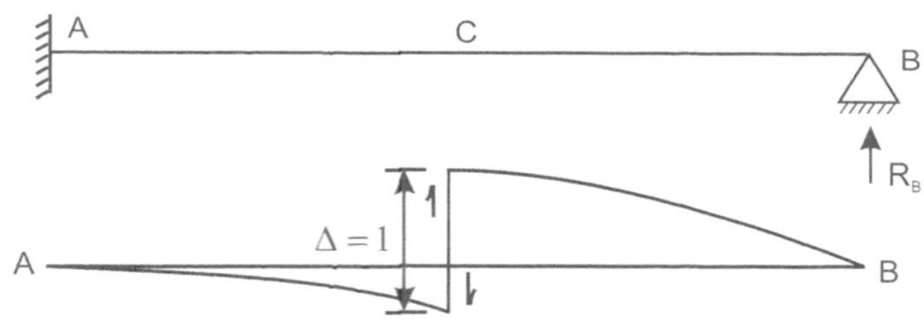

Fig. 5. ILD for SF_C.

8. Sketch the influence line diagram for moment at the fixed end of a propped cantilever using Muller-Breslau principle.

Ans. Introduce a hinge at A (remove the constraint to moment at A). Apply sufficient moment at A in order to produce a unit rotation of the beam at A. The resulting deflected line is the influence line diagram for moment at A (see Fig. 6).

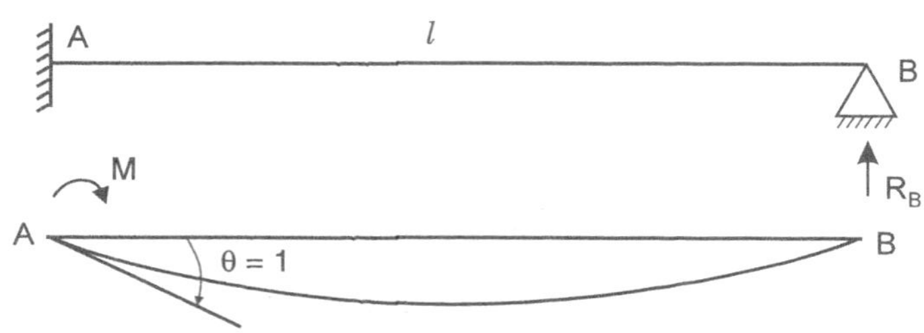

Fig. 6. ILD for M_A.

9. Draw the qualitative influence line diagrams for the reactions of a fixed beam.

Ans. Provide rollers at the fixed end A as shown in Fig. 7 (*b*). This would ensure that end A can move up or down but would not permit change of slope at A.

Now apply a unit vertical displacement at A (corresponding to reaction R_A). The resulting elastic curve is the influence line diagram for R_A. Likewise, we can sketch the influence line diagram for R_B [see Fig. 7 (*c*) and (*d*)].

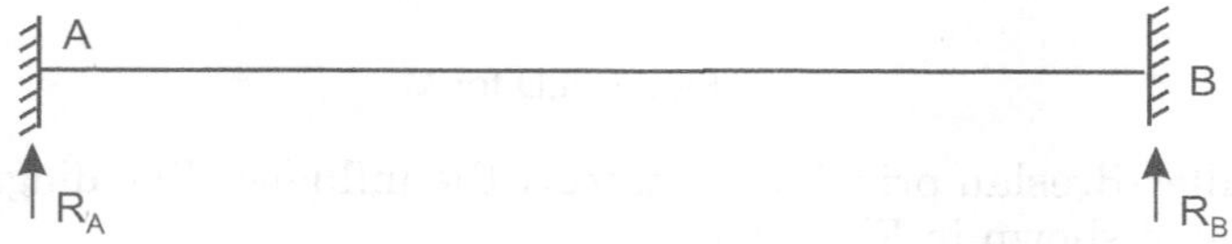

Fig. 7 (a)

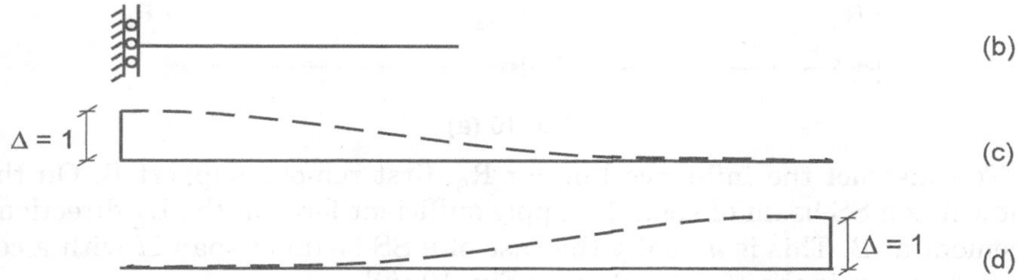

Fig. 7. (b) rollers ; (c) ILD for R_A ; (d) ILD for R_B.

10. Sketch the influence line diagram for shear force at a section X on a fixed beam.

Ans. Introduce at X a cut in the beam and provide a sliding joint there. This would permit up and down movement between the two parts of the beam. But the slope would be the same to the left and right of the section X (The resistance to bending moment would be intact).

Now give a unit shear displacement between the 2 parts of the beam at X. The resulting elastic curve is the influence line diagram for shear at X (see Fig. 8).

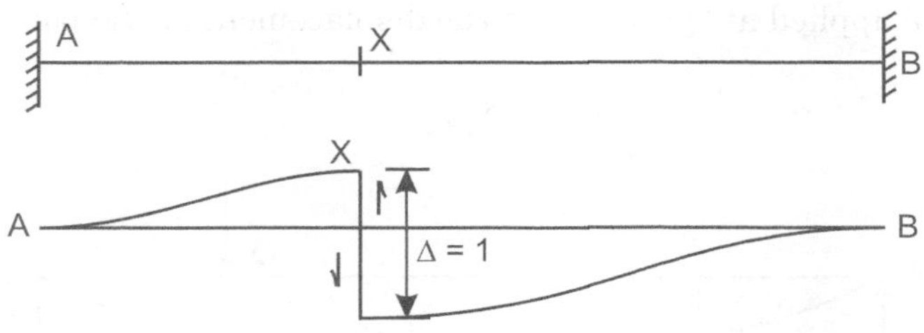

Fig. 8. ILD for SF_x.

11. Sketch the influence line diagram for bending moment at a section X of a fixed beam.

Ans. Introduce a hinge at X (remove the restraint to bending moment). Apply a suitable bending moment at X so as to produce a unit angular rotational displacement between the left part of the hinge and the right part. The resulting elastic curve is the influence line diagram for bending moment at X (see Fig. 9).

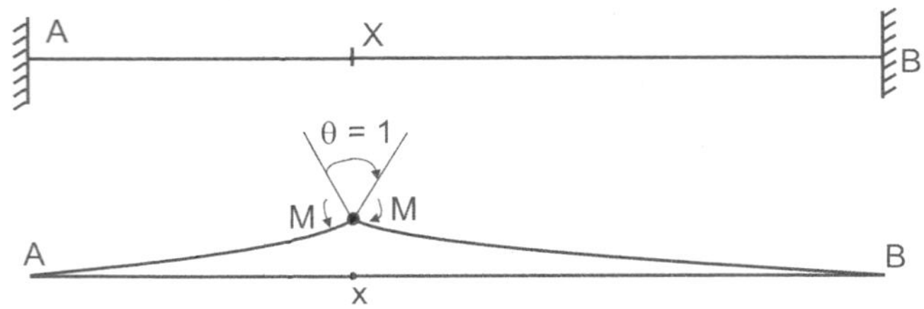

Fig. 9. ILD for M_x.

12. Using Muller-Breslau principle, construct the influence line diagram for R_B in the continuous beam shown in Fig. 10 (*a*).

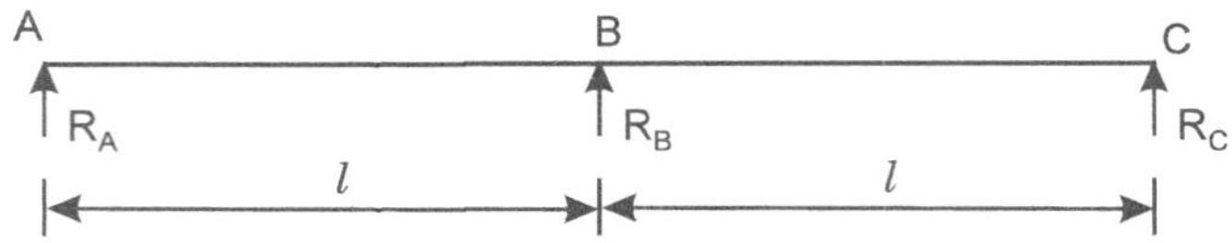

Fig. 10 (a)

Ans. To construct the influence line for R_B, first remove support B. On this reduced structure (now it is a SS beam of span $2l$), apply sufficient force in the R_B direction to cause a unit displacement at B. This is actually the case of a SS beam of span $2l$ with a central load. The deflection diagram is the ILD for R_B [see Fig. 10 (*b*)].

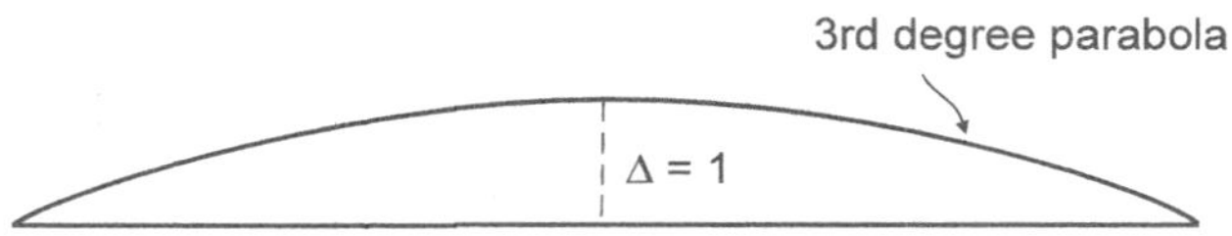

Fig. 10. (b) ILD for R_B.

13. State Maxwell's theorem of reciprocal deflection.

Ans. When a linearly elastic structure is in static equilibrium, the displacement at a point P due to a force applied at Q is equal to the displacement at the point Q due to the same force applied at P.

i.e., $$\delta_{PQ} = \delta_{QP}$$

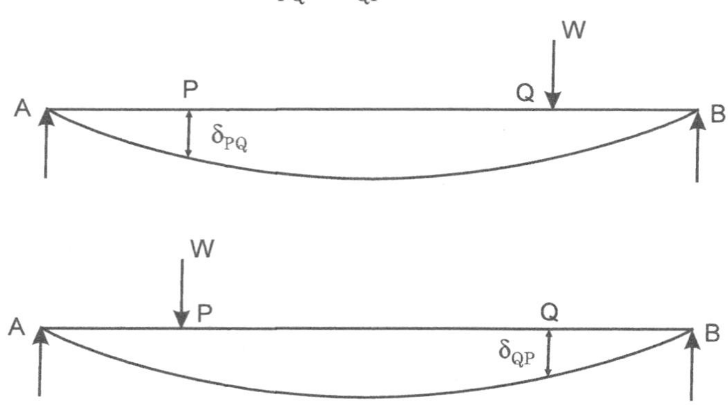

Fig. 11

14. Sketch the influence line diagram for horizontal reaction at the left support of a single bay single storey portal with fixed legs.

Ans. To draw the influence line diagram for H_A, place support A on rollers so that horizontal displacement at A is permitted. Apply a unit horizontal displacement at A. The resulting displaced shape is the ILD for H_A.

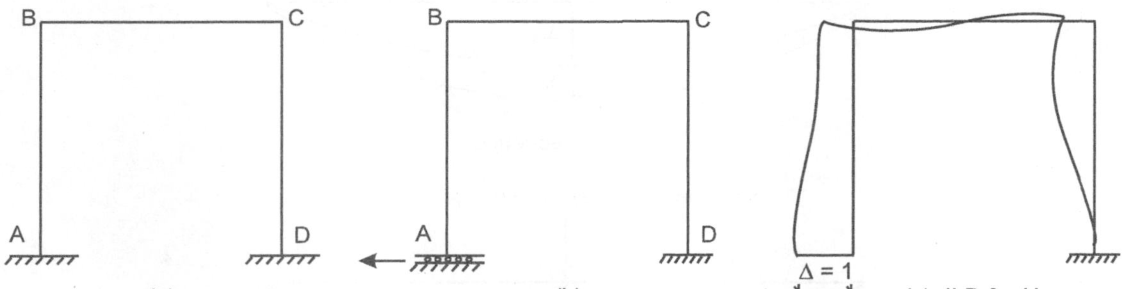

(a) (b) 'A' placed on rollers (Δ_H is permitted; but slope is arrested) (c) ILD for H_A

Fig. 12

15. Demonstrate Betti's theorem by considering a SS beam with a load at mid-point and the same beam with a load at a quarter point.

Ans.

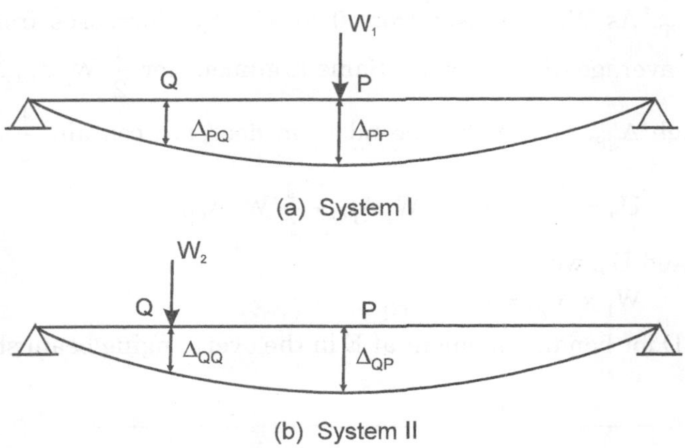

Fig. 13. (a) System I, (b) System II

According to Betti's theorem,

$$W_1 \times D_{QP} = W_2 \times D_{PQ}$$

We can demonstrate this by,

1. Applying W_1 first and W_2 next and getting the work done U_1.
2. Applying W_2 first and W_1 next and getting the work done U_2.

In a linear elastic structure, the work done by loads shall not depend on the sequence of loading. Hence $U_1 = U_2$.

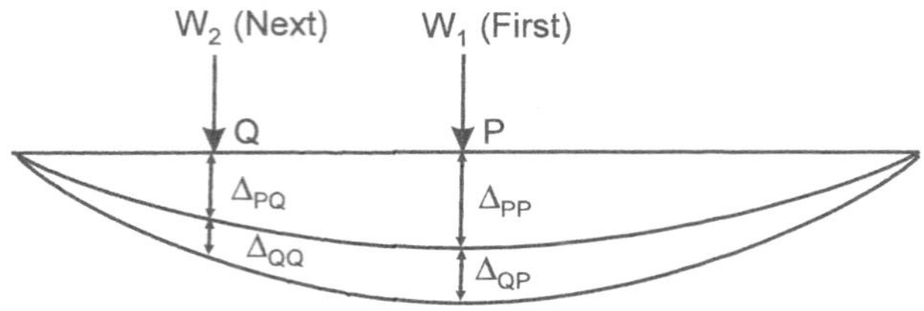

(c) First sequence

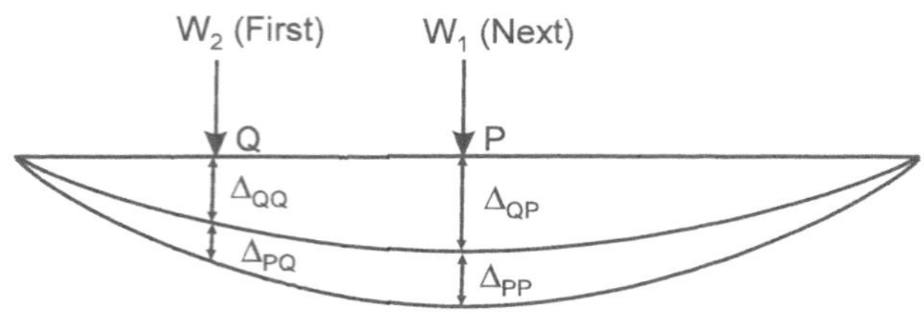

(d) Second sequence

Fig. 13. (c) First sequence, (d) Second sequence

$$U_1 = ½\, W_1\, \Delta_{PP} + W_1\, \Delta_{QP} + ½\, W_2\, \Delta_{QQ}$$

(W_1 causes Δ_{PP}. As W_1 increases from 0 to W_1, Δ_{PP} increases from 0 to Δ_{PP}. Hence work = final force × average displacement. Same argument for $\frac{1}{2} W_2\, \Delta_{QQ}$. But the full value of W_1 travels through Δ_{QP}. Hence the second term does not contain $\frac{1}{2}$).

$$U_2 = \frac{1}{2} W_2\, \Delta_{QQ} + W_2\, \Delta_{PQ} + \frac{1}{2} W_1\, \Delta_{PP}$$

Equating U_1 and U_2, we get

$$W_1 \times \Delta_{QP} = W_2 \times \Delta_{PQ}$$

16. Sketch the ILD for bending moment at E in the overhanging beam shown in Fig. 14 (*a*).

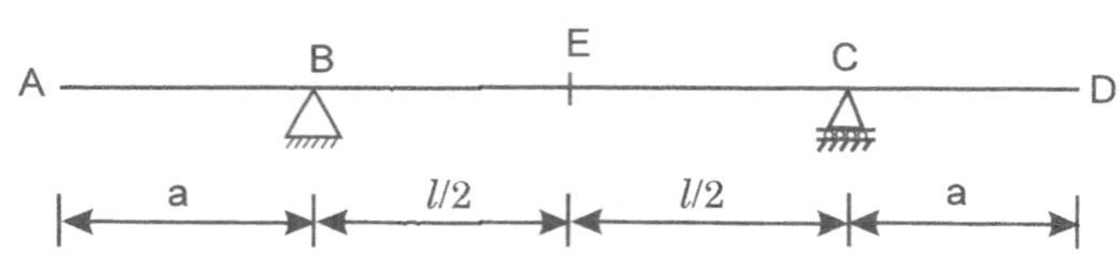

Fig. 14 (a)

Ans. We will use Muller-Breslau principle. First introduce a hinge at E. (Remove the resistance to bending moment). Then apply a pair of moments at E so as to create a unit change of slope between EB and EC (or their normals). With the hinge E in place, the structure is deficient and the moment required to cause this unit rotation is zero. The displaced configuration is the influence line for M_E [Fig. 14 (*b*)].

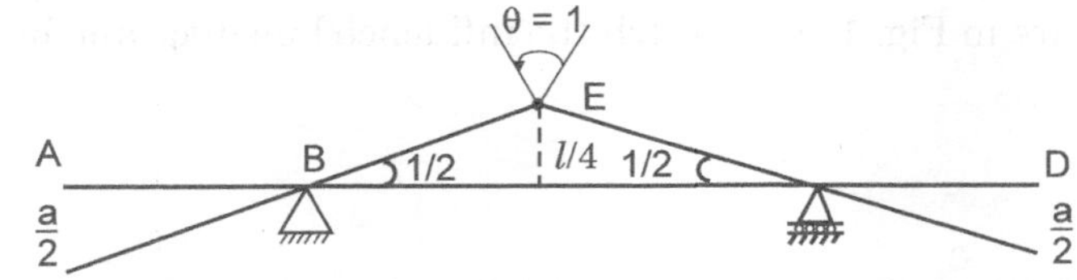

Fig. 14. (b) ILD for M_E.

17. Sketch the influence line for reaction at B in the overhanging beam shown in Fig. 15 (*a*).

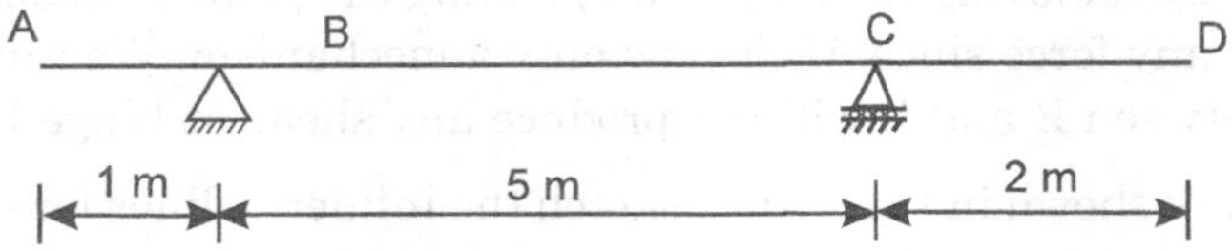

Fig. 15 (a)

Ans.

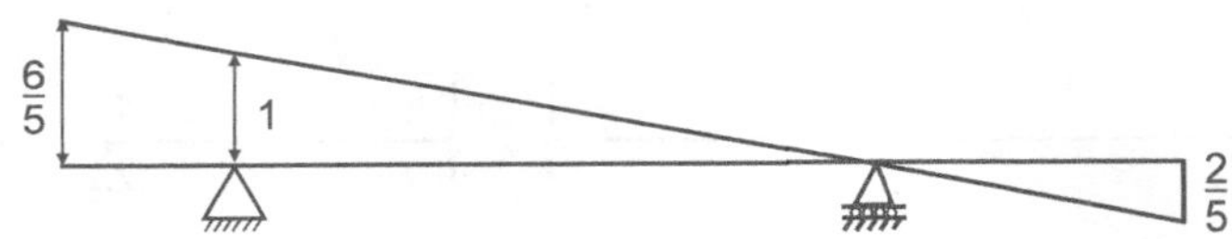

Fig. 15. (b) ILD for R_B.

18. A balanced cantilever bridge is articulated so as to remain determinate. Sketch the influence line diagram for the hinge reaction at C in Fig. 16 (*a*).

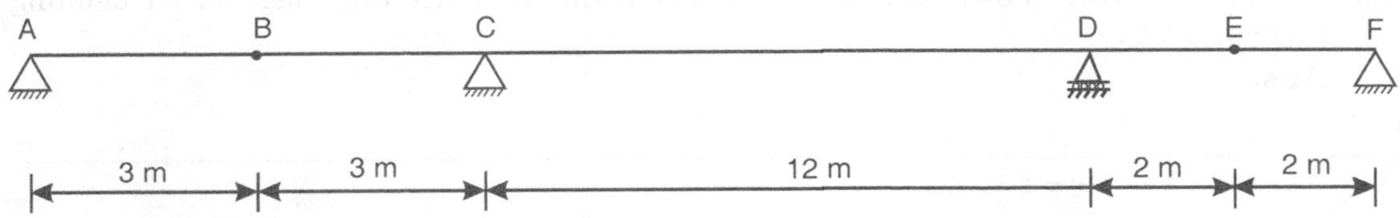

Fig. 16. (a) Balanced cantilever bridge.

Ans.

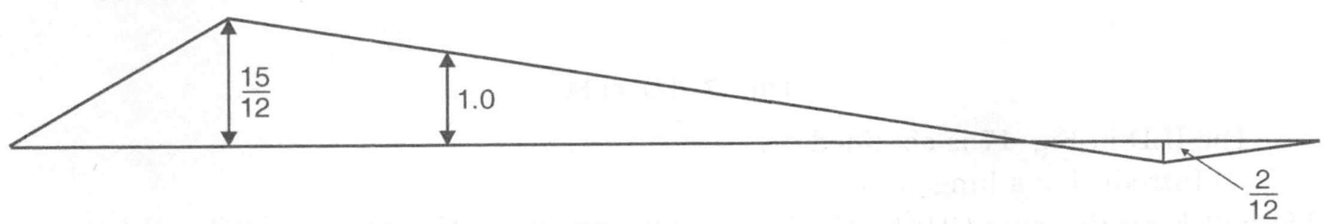

Fig. 16. (b) ILD for R_C.

The ILD in Fig. 16 (*b*) is obtained by

(*i*) Removing the support at C.

(*ii*) Applying a unit upward displacement at C (corresponding to reaction R_C)

The resulting (rigid body) displaced centre line is the influence line diagram for R_C.

19. For the structure in Fig. 16 (*a*), sketch the influence line diagram for shear at the hinge B (SF_B).

Ans.

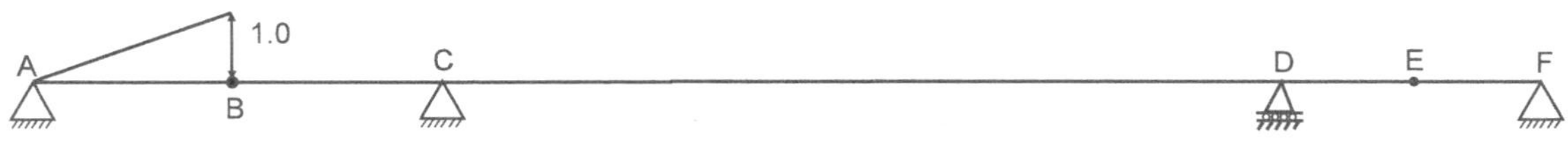

Fig. 16. (c) ILD for SF_B.

The ILD is got by removing the hinge and pushing the point B (of span AB) up by 1 unit. This will not require any force since AB is now only a mechanism. We can infer that any load applied anywhere between B and F will not produce any shear at hinge B.

20. For the structure shown in Fig. 16 (*a*), sketch the influence lines for reactions at A and F.

Ans. It is obvious that A will develop a reaction only if there is a load on AB. Likewise, only a load on EF can produce a reaction at F. Hence the ILDs in Fig. 16 (*d*).

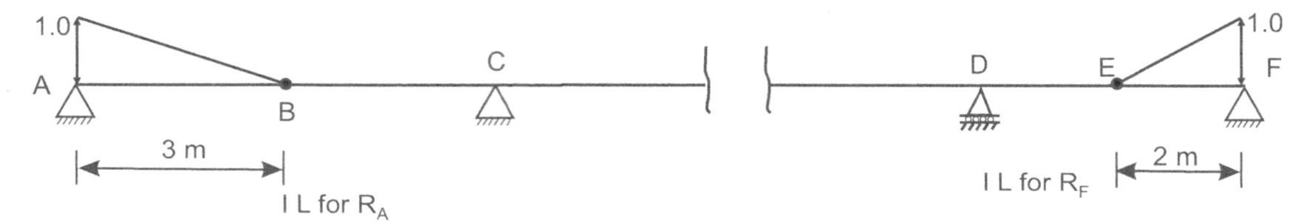

Fig. 16 (d)

21. For the structure shown in Fig. 16 (*a*), sketch the influence line diagram for bending moment at C.

Ans.

Fig. 17. ILD for M_C.

The ILD in Fig. 17 is obtained by,

(*i*) Introducing a hinge at C.

(This will keep the part CDEF still as a stable structure. The portion ABC will become a mechanism).

(*ii*) Applying at C a unit rotation between CB and CD.

The rotation will be only on CB. Hence, the deflected shape in Fig. 17 is the ILD for M_C.

22. For the 2 member determinate bent in Fig. 18 (*a*), sketch the influence line for V_A.

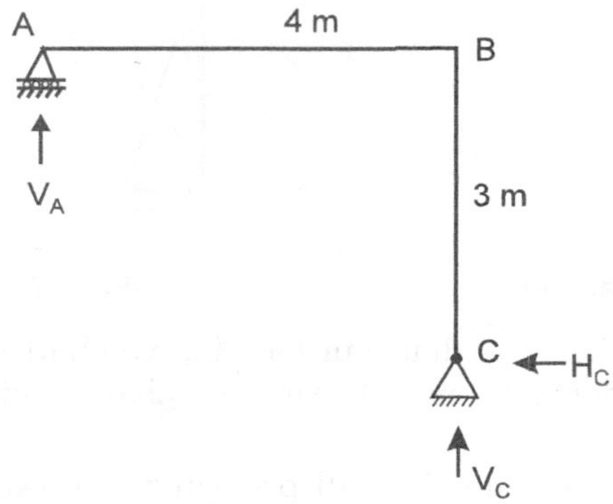

Fig. 18 (a)

Ans. If we deftly apply Muller-Breslau to the problem, we can first remove support A and push A up by unit distance. Since the angle at B will remain unchanged, our unit displacement will result in a rigid body rotation of θ (= ¼ radian) about C. So the column CB will also have a horizontal displacement (for the part AB, the diagram is just the influence line diagram for shear in a S.S. beam).

The diagram for BC must be understood to be the influence of horizontal loads on the column on R_A.

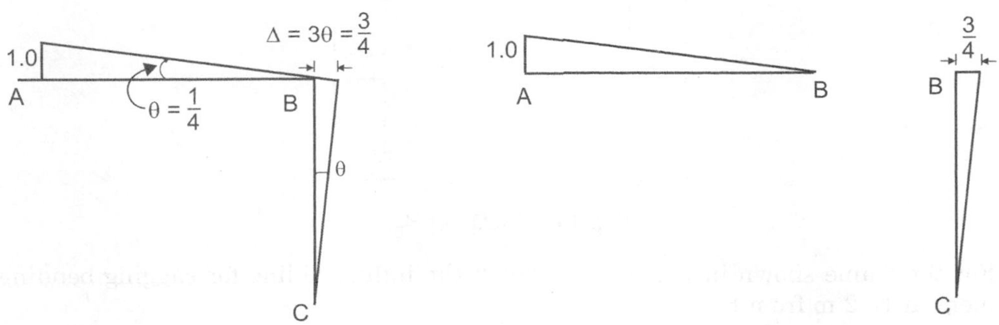

Fig. 18. (b) ILD for V_A.

23. The influence line for vertical reaction in a frame shown in Fig. 18 (*a*) is given in Fig. 18 (*b*). Find the reaction at A when a uniformly distributed horizontal force of 3 kN/m acts on column BC from right to left.

Ans. R_A = load intensity × area of ILD on BC

$$R_A = 3 \times \frac{1}{2} \times \frac{3}{4} \times 3 = \frac{27}{8} = 3.375 \text{ kN.}$$

24. The portal frame in Fig. 19 (*a*) is hinged at D and is on rollers at A. Sketch the influence line for bending moment at B.

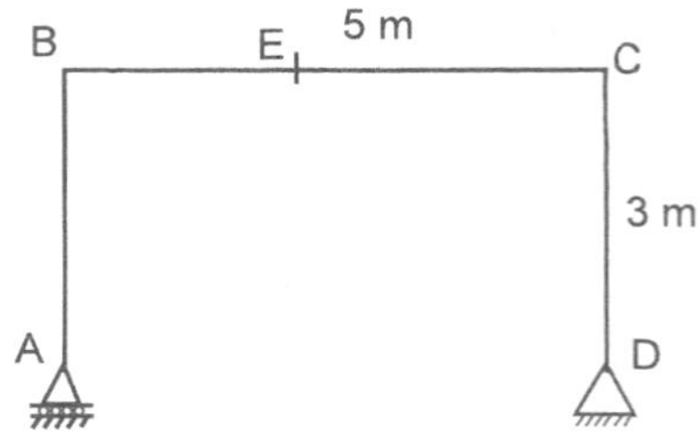

Fig. 19. (a) Portal frame.

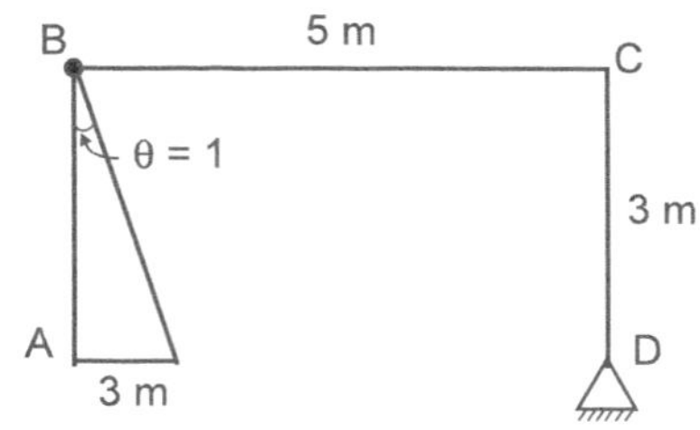

Fig. 19. (b) ILD for M_B.

Ans. To get the influence line diagram for M_B, we shall introduce a hinge at B (and remove the resistance to bending moment). Now we give a unit rotation between BA and BC at B.

BC cannot rotate since column CD will prevent the rotation. BA would rotate freely (with zero moment). For $\theta = 1$ at B, displacement at A = 3 m. The displaced position shows the influence line for M_B, as shown in Fig. 19 (*b*).

25. For the frame shown in Fig. 19 (*a*). Sketch the influence line diagram for horizontal reaction at D.

Ans. To find the influence line for H_D, remove the horizontal restraint at D (keep D on rollers). Now push D by a unit horizontal displacement towards A. The whole frame will now make a rigid body horizontal displacement [Fig. 19 (*c*)]. This displacement diagram is also the influence line for H_D.

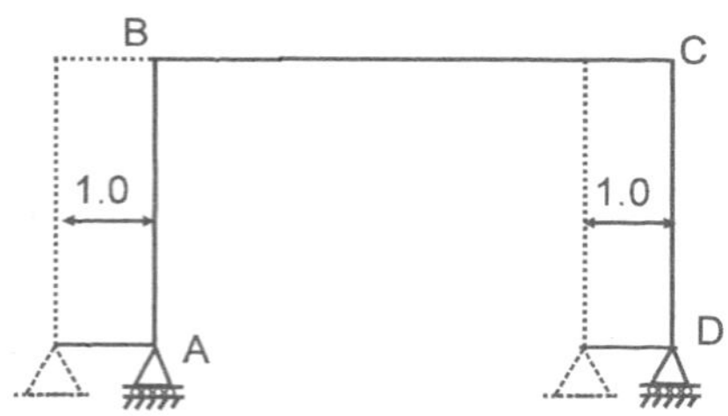

Fig. 19. (c) ILD for H_D.

26. For the frame shown in Fig. 19 (*a*), sketch the influence line for sagging bending moment at E, 2 m from B.

Ans. First, introduce a hinge at E (thus removing the restraint to bending moment). Introduce an angular displacement of 1 radian between the normals to BE and CE. The resulting displacement diagram is the influence line diagram for M_E [see Fig. 19 (*d*)].

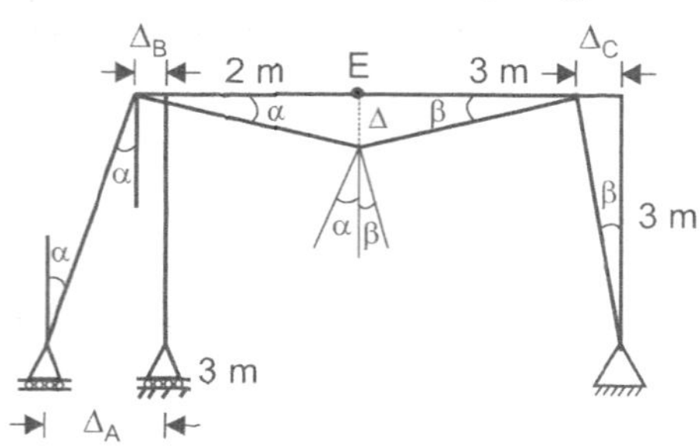

Fig. 19 (d)

When we apply a unit rotation between EB and EC, we shall work out the kinematics.

$$\Delta = 2\alpha = 3\beta \qquad \therefore \quad \alpha = 1.5\beta$$
$$\alpha + \beta = 1\ ; \quad 2.5\beta = 1\ ; \quad \beta = 0.4\ ; \quad \alpha = 0.6$$
$$\Delta_C = 3\beta = 3 \times 0.4 = 1.2$$
$$\Delta_B = \Delta_C = 1.2$$
$$\Delta_A = 3.0 \qquad (\because\ \Delta_A = \Delta_B + 3\alpha = 1.2 + 3 \times 0.6 = 3.0)$$
$$\Delta = 2\alpha = 1.2$$

Thus the influence line diagram can be drawn as shown in Fig. 20.

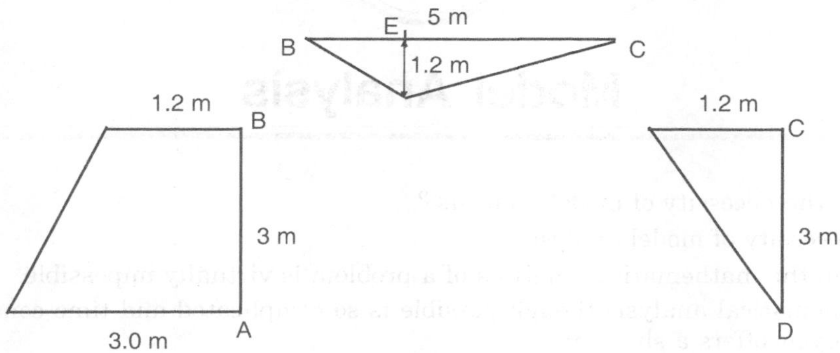

Fig. 20. IL for M_E.

27. For the bent cantilever shown in Fig. 21 (*a*), sketch the influence line for bending moment at D.

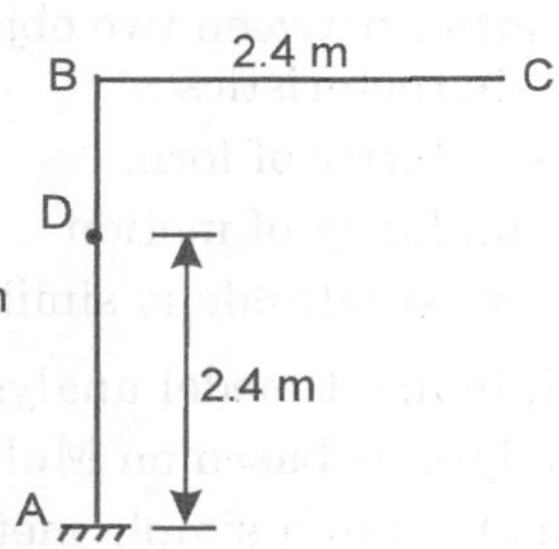

Fig. 21. (a) Cantilever bent.

Ans. Introduce a hinge at D. Rotate DBC by 1 radian anticlockwise. The resulting displaced line is the influence line diagram for M_D [see Fig. 21 (*b*)].

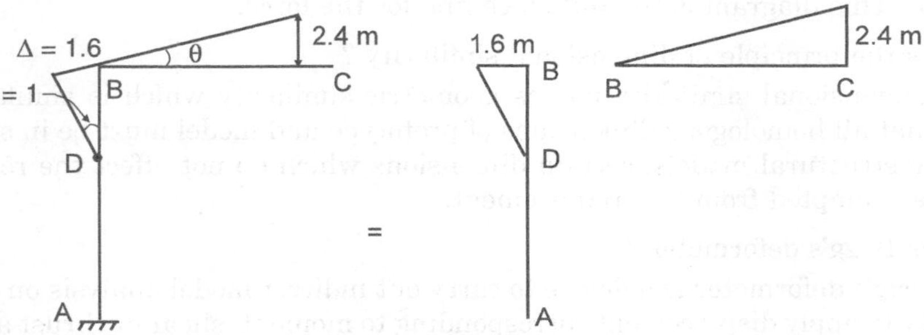

Fig. 21. (b) ILD for M_D.

12

Model Analysis

1. What is the necessity of model analysis ?

Ans. Necessity of model analysis :

(*i*) When the mathematical analysis of a problem is virtually impossible.

(*ii*) Mathematical analysis though possible is so complicated and time consuming that the model analysis offers a short-cut.

(*iii*) The importance of the problem is such that verification of mathematical analysis by an actual test is essential.

2. Define 'similitude'.

Ans. Similitude means similarity between two objects namely the model and the prototype with regard to their physical characteristics :

- Geometric similitude is similarity of form
- Kinematic similitude is similarity of motion
- Dynamic and/or mechanical similitude is similarity of masses and/or forces.

3. State the principle on which indirect model analysis is based ?

Ans. The indirect model analysis is based on Muller Breslau principle.

Muller Breslau principle has lead to a simple method of using models of structures to get the influence lines for force quantities like bending moments, support moments, reactions, internal shears, thrusts, etc.

To get the influence line for any force quantity, (*i*) remove the restraint due to the force, (*ii*) apply a unit displacement in the direction of the force, and (*iii*) plot the resulting displacement diagram. This diagram is the influence line for the force.

4. What is the principle of dimensional similarity ?

Ans. Dimensional similarity means geometric similarity which is similarity of form. This means that all homologous dimensions of prototype and model must be in some constant ratio. In some structural models, certain dimensions which do not affect the response of the model may be exempted from this requirement.

5. What is Begg's deformeter ?

Ans. Begg's deformeter is a device to carry out indirect model analysis on structures. It has the facility to apply displacement corresponding to moment, shear or thrust at any desired point in the model. In addition, it provides facility to measure accurately the consequent displacements all over the model.

6. Give the dimensions of : (*a*) acceleration and (*b*) pressure gradient.

Ans. Acceleration — LT^{-2}

Pressure gradient — FL^{-3}, where F is M LT^{-2}

7. What are scale factors ?

Ans. As applied to statically loaded structural prototypes and models, the requirements for similitude are:

(*i*) all homologous linear dimensions must be in some constant ratio, α

(*ii*) all homologous linear displacements must be in some constant ratio, β

(*iii*) all homologous external loads, internal elastic forces and dead loads must be in some constant ratio, γ

These constant ratios α, β and γ are known as scale factors. α, β and γ may be inter related.

8. List any four desirable characteristics of model making materials.

Ans. The model making material should be : inexpensive; easily obtainable; easy to cut, shape and joint together; physical properties should not change with time, humidity, stress or temperature; should obey Hooke's law and should be isotropic.

9. Name any 4 model making materials.

Ans. Perspex, Plexiglas, acrylic, plywood, sheet araldite and bakelite are some of the model making materials. Micro-concrete, mortar and plaster of paris can also be used for models.

10. A 24 m span psc girder is to be modelled for indirect model analysis. The moments of inertia of the girder are as under. Design a model and sketch the dimensioned profile.

Section	1	2	3	4	5
M.I. $cm^4 \times 10^{-6}$	0.96	0.64	0.36	0.30	0.28

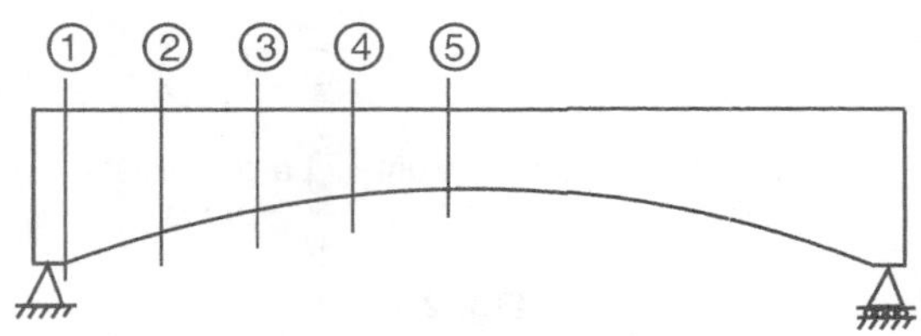

Fig. 1

Ans. The model will be made of some sheet plastic like perspex.

Let the thickness of the model be 1.2 mm. Let us have the maximum width of the model as 15 mm (to fit into the Begg's deformeter gauge)

Then d_1 = 15 mm and its M.I. should be proportional to 0.96×10^6 cm^4.

$$\frac{1}{12} \times 1.2 \times 15^3 = k \times 0.96 \times 10^6$$

where k is the scale for M.I.

$$k = \frac{15^3}{10 \times 0.96 \times 10^6} = \frac{1}{2844}$$

For section 2, $\frac{d_2^{\,3}}{10} = k \times 0.64 \times 10^6$ $\qquad d_2 = 13.1$ mm

For section 3, $\frac{d_3^{\,3}}{10} = k \times 0.36 \times 10^6$ $\qquad d_3 = 10.8$ mm

For section 4, $\frac{d_4^{\,3}}{10} = k \times 0.30 \times 10^6$ $\qquad d_4 = 10.2$ mm

For section 5, $\frac{d_5^{\,3}}{10} = k \times 0.28 \times 10^6$ $\qquad d_5 = 9.9$ mm

11. What is 'dummy length' in models tested with Begg's deformeter.

Ans. Dummy length is the additional length (of about 10 to 12 mm) left at the extremities of the model to enable any desired connection to be made with the gauges.

12. What are the three types of connections possible with the model used with Begg's deformeter.

Ans. 1. Hinged connection
2. Fixed connection and
3. Floating connection

are the three types of connections possible.

13. What is the use of a micrometer microscope in model analysis with Begg's deformeter?

Ans. Micrometer microscope is an instrument used to measure the displacements of any point in the x and y directions of a model during tests with Begg's deformeter.

14. Sketch the plugs used for inducing axial displacements in the model analysis using Begg's deformeter.

Ans.

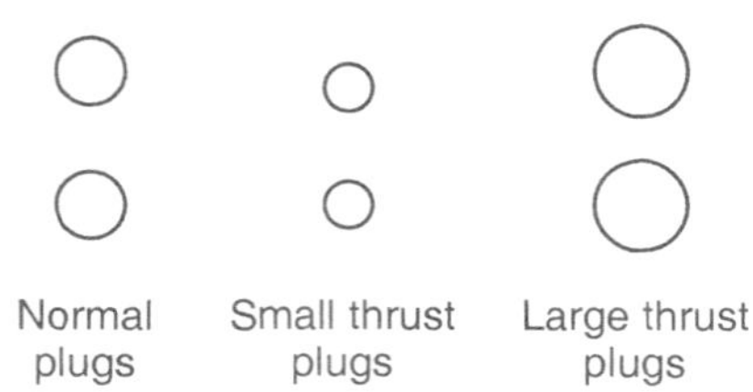

Fig. 2

Fig. 2 shows the plugs used for inducing axial displacement in models using Begg's deformeter.

15. Sketch the shear plugs and moment plugs used in Begg's deformeter.

Ans.

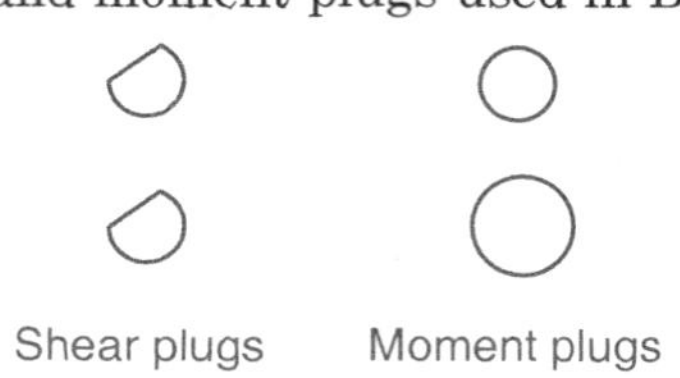

Fig. 3

16. In modelling a gravity dam, a linear scale of 1 : 100 was chosen. Specific gravity of the dam materials was 2.1. If the specific gravity of the model material is 7.8, what would be the scale obtained for self weight forces ?

Ans. Linear scale $= 1/100 = 10^{-2}$

Volume scale $= 10^{-6}$

Weight scale $= 10^{-6} \times \dfrac{7.8}{2.1} = 3.71 \times 10^{-6}$.

17. In question 16, suggest a method of simulating self weight suitably.

Ans. A scale of 3.71×10^{-6} is much too small for measurement of forces on the model. So, body forces will have to be simulated

(*i*) by using external applied loading

(*ii*) by using a magnetic material for the model and applying a magnetic field to induce body force in the gravity direction.

18. The deflection of a truss under loading is expressible in the form $\Delta = \Sigma \dfrac{Ful}{AE}$.

The variables being,

(*i*) linear dimensions, l

(*ii*) deflections, Δ

(*iii*) loads, W

(*iv*) member forces, F

(*v*) forces per unit load, u

(*vi*) areas of cross-section, A and

(*vii*) Young's modulus, E

Assuming that these are expressible in terms of force and length units alone, suggest suitable π terms to describe the phenomenon of deflection of a truss.

Ans. According to Buckingham π theorem, p, the number of π terms $= n-m$ where, in this case $n = 7$, the number of variables

$m = 2$, the number of dimensions (force, length)

$$\pi = 7 - 2 = 5$$

$$\Delta = \Sigma \frac{Ful}{AE}$$

The π terms are

$$\pi_1 = \frac{F}{AE}; \quad \pi_2 = \frac{A}{l^2}; \quad \pi_3 = \frac{\Delta}{l}; \quad \pi_4 = u = \frac{F}{W}; \quad \pi_5 = \frac{W}{AE}$$

π_1, π_2, π_3, π_4 and π_5 will have the same values for the model and the true structure.

19. The deflections in a beam is expressible as $\delta = \alpha \dfrac{Wl^3}{EI} + \beta \dfrac{wl^4}{EI}$, where W are concentrated loads and w are distributed loads.

If we want to carry out a model analysis of the phenomenon, choose the appropriate dimensionless π terms.

Ans. The given phenomenon can be written as $f(\delta, W, l, E, I, w) = 0$.

There are 6 variables.

These are all expressible in terms of 2 basic dimensions, force W and length l.

We can therefore, chose $6 - 2 = 4$ π terms.

$$\pi_1 = \frac{\delta}{l}\ ; \qquad \pi_2 = \frac{El^2}{W}\ ; \qquad \pi_3 = \frac{I}{l^4}\ ; \qquad \pi_4 = \frac{wl}{W}$$

π_1, π_2, π_3 and π_4 will be the same for the model and the real structure.

20. A truss develops forces F_1, F_2, F_3 in its 3 members due to an applied load W at C. The members have sectional areas a_1, a_2 and a_3. Design a 1 : 10 linear scale model and choose π terms so that they are the same for the model and the structure it simulates.

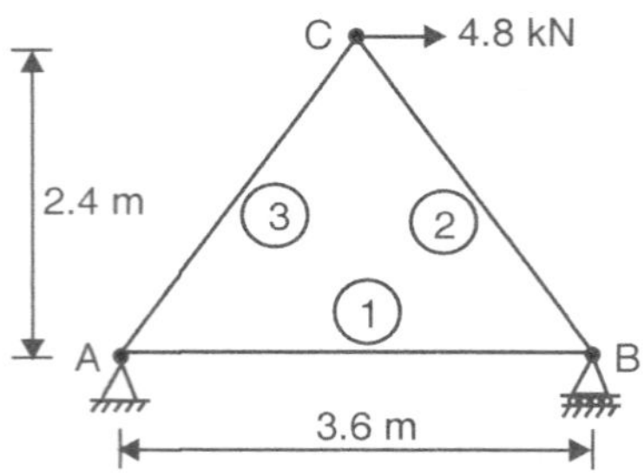

Fig. 4

Ans. Let us assume that the following are the variables in the analysis.

F_1, F_2, F_3, span l, height h, load W, reactions V_A and V_B and stresses σ_1, σ_2 and σ_3.

All the 11 variables above are expressible in terms of force dimension W and length dimension l.

Hence, we have 11 – 2 = 9 dimensionless terms called π terms. They are

$$\pi_1 = \frac{F_1}{W}\ ; \qquad \pi_2 = \frac{F_2}{W}\ ; \qquad \pi_3 = \frac{F_3}{W}\ ; \qquad \pi_4 = \frac{h}{l}\ ;$$

$$\pi_5 = \frac{V_A}{W}\ ; \qquad \pi_6 = \frac{V_B}{W}\ ; \qquad \pi_7 = \frac{\sigma_1 l^2}{W}\ ; \qquad \pi_8 = \frac{\sigma_2 l^2}{W}\ ; \qquad \pi_9 = \frac{\sigma_3 l^2}{W}$$

Once we have found the π terms for the model, the same are valid for the true structure.

21. In question 20, if scale n = 10 and if $\pi_7, \left(= \frac{\sigma_1 ll}{W}\right)$ is found to be 244 for the 1 : 10 model, find σ_1 for the structure shown in Fig. 4.

Ans. For the structure, W = 4.8 kN, l = 3.6 m

$$\frac{\sigma_1 ll}{W} = 244$$

$$\sigma_1 = \frac{244 \times 4.8}{3.6 \times 3.6} = 90.37$$

22. What load should be applied on the scaled down model (1 : 10) to simulate the loaded structure?

Ans. The similitude is valid for all loads. No scaling down of the load is needed.

23. An I section has a depth of 300 mm and M.I. of 72 × 10^6 mm^4. Its scaled down model is of aluminium, depth 30 mm and M.I. 7200 mm^4. Work out the strain in the model corresponding to a flexural strain of 30×10^{-6} in the structure. [E for steel = 200 kN/mm^2, E for aluminium is 80 kN/mm^2].

Ans. $$\varepsilon = \frac{\sigma}{E} = \frac{Md}{2IE}$$

Here E value can not be scaled down at will.

$$\frac{E_s}{E_m} = \frac{200}{80} = 2.5$$

The other variables are σ, M, d and I.

The basic units are F and L.

There will be two π terms

$$\pi_1 = \frac{\sigma}{M} d^3$$

$$\pi_2 = \frac{I}{d^4}$$

Now π_1 term is same for model and structure.

So, π_1, the stress term in model = stress term in structure

$$\frac{\sigma_m d_m^{\ 3}}{M_m} = \frac{\sigma_s d_s^{\ 3}}{M_s}$$

$$\sigma_m = \sigma_s \frac{d_s^{\ 3}}{d_m^{\ 3}} \frac{M_m}{M_s}$$

If we chose linear scale as 1 : 10 and moment scale as 1 in 4000.

$$\sigma_m = \sigma_s (10)^3 \div 4000 = \sigma_s \times 0.25$$

$$\varepsilon_m = \frac{\sigma_m}{E_m} = \frac{\sigma_s \times 0.25}{E_m} = \frac{\sigma_s}{E_s} \frac{E_s}{E_m} \times 0.25$$

$$\varepsilon_m = \varepsilon_s \times 2.5 \times 0.25 = 0.625\ \varepsilon_s.$$

24. It is proposed to analyse a R.C. arch bridge of span 32 m by indirect model analysis. The maximum rise of the bridge is 6 m. The size of the drawing board available for testing is 600 mm × 420 mm. Determine suitable dimensions for the model.

Ans. Let us make a model 50 cm long

Linear scale of the model $= \dfrac{3200 \text{ cm}}{50 \text{ cm}} = 64$

Vertical dimension of the model $= \dfrac{600 \text{ cm}}{64} = 9.375$ cm.

13

Arches

1. What is an arch ? Explain.

Ans. An arch is defined as a curved girder, having convexity upwards and supported at its ends.

The supports must effectively arrest displacements in the vertical and horizontal directions. Only then there will be arch action.

2. What is a linear arch ?

Ans. If an arch is to take loads, say W_1, W_2 and W_3 (Fig. 1) and a vector diagram and funicular polygon are plotted as shown, the funicular polygon is known as the linear arch or theoretical arch.

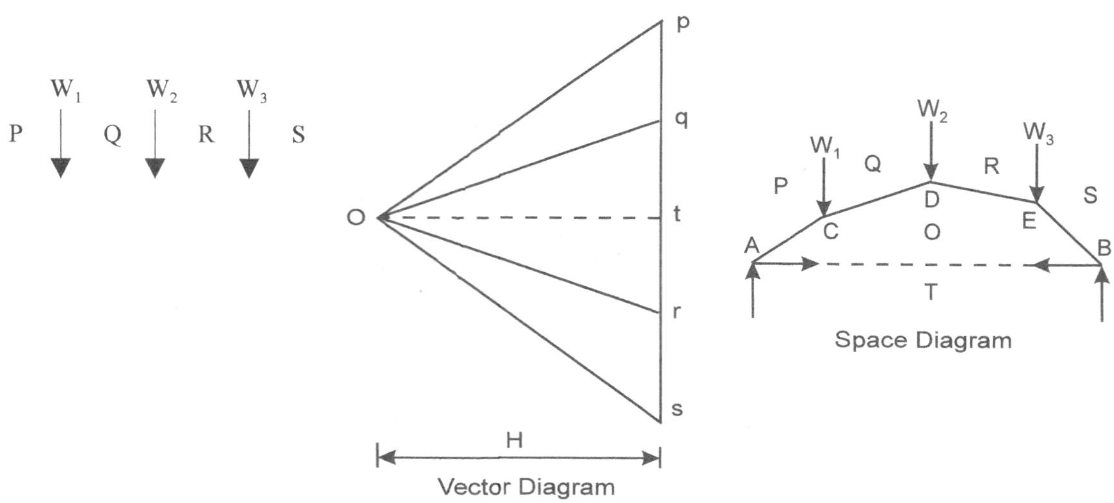

Fig. 1

The polar distance 'ot' represents the horizontal thrust.

The links AC, CD, DE and EB will be under compression and there will be no bending moment. If an arch of this shape ACDEB is provided, there will be no bending moment.

For a given set of vertical loads W_1, W_2, ... etc., we can have any number of linear arches depending on where we choose 'O', or how much horizontal thrust (ot) we choose to introduce.

3. State Eddy's theorem.

Ans. Eddy's theorem states that "The bending moment at any section of an arch is proportional to the vertical intercept between the linear arch (or theoretical arch) and the contre line of the actual arch".

$$BM_x = \text{ordinate } O_2O_3 \times \text{scale factor}$$

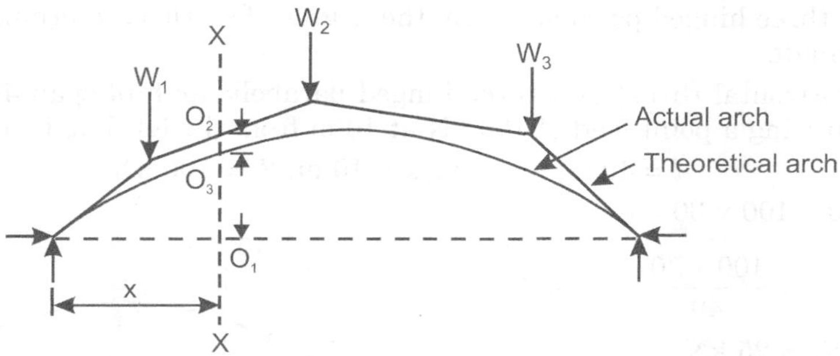

Fig. 2

4. Define the structural aspect of arch.

Ans. Arches are shaped to take the load above them and develop only compression. However, they are predominantly in compression and do develop bending moments and shear too because of its rigidity and depending upon how it is loaded and shaped.

5. Comment on the role of "hinges" in arches.

Ans. It has been the practice to introduce structural hinges in arches. Hinges save the computational effort but, make the arches structurally less efficient. The justification for hinges at the springings is that the supports (piers or abutments) are kept free of any moments transmitted from the arches.

6. List out the different types of arches.

Ans. Arches can be built of masonry, reinforced concrete or steel.

Based on shape, arches can be:

1. Polygonal 2. Circular 3. Parabolic
4. Elliptical and 5. Of any other curved shape.

Based on the hinges provided, arches can be:

1. Three hinged 2. Two hinged and 3. Fixed.

7. Define the term "Horizontal thrust".

Ans. The external load on the arch will cause a sagging bending moment in the arch which will try to make the arch flat. At the hinges at both springings, a horizontal force, H will be developed which will help the arch to have arch action. The horizontal force will cause hogging bending moment and also axial force component due to H. Such a horizontal force developed at the springings is called the horizontal thrust, H.

8. What are the three forces acting at any section of an arch ?

Ans. Any section of an arch will be subjected to:

1. Normal thrust, P or N
2. Radial shear, F or R and
3. Bending moment, M.

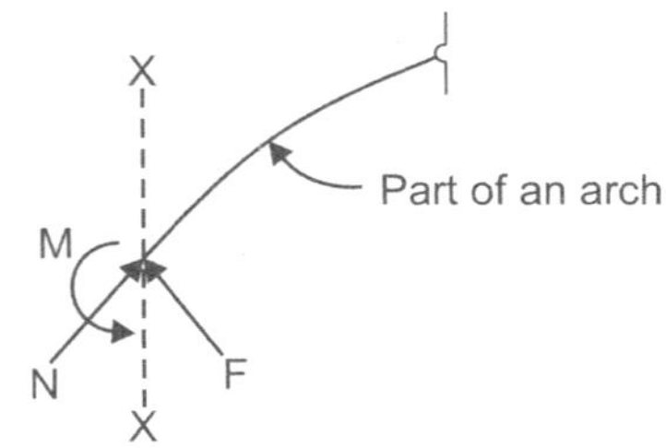

Fig. 3

9. What is the degree of static indeterminacy of a three hinged parabolic arch ?

Ans. For a three hinged parabolic arch, the degree of static indeterminacy is zero. It is statically determinate.

10. Find the horizontal thrust of a three hinged parabolic arch of span 40 m and central rise 6 m carrying a point load of 100 kN at 10 m from the left hand support.

Ans. $l = 40$ m, $y_c = 6$ m, $x = 10$ m, W = 100 kN

$$V_A \times 40 - 100 \times 30 = 0$$

$$V_A = \frac{100 \times 30}{40} = 75 \text{ kN}$$

$$V_B = 25 \text{ kN}$$

$$M_C = V_B \times 20 - H.y_c = 0$$

$$25 \times 20 = H \times 6$$

$$\text{Horizontal thrust, } H = \frac{25 \times 20}{6} = 83.33 \text{ kN}$$

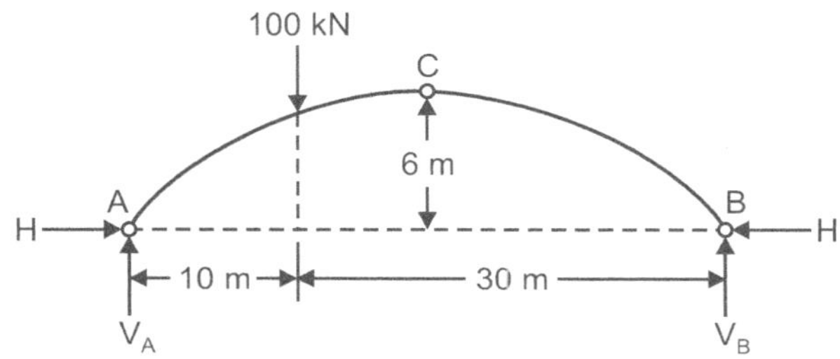

Fig. 4

11. A three hinged parabolic arch hinged at the crown and springings has a horizontal span of 12 m and a central rise of 2.5 m. It carries a udl of 30 kN/m run over the left hand half of the span. Calculate the reactions at the end hinges.

Ans.

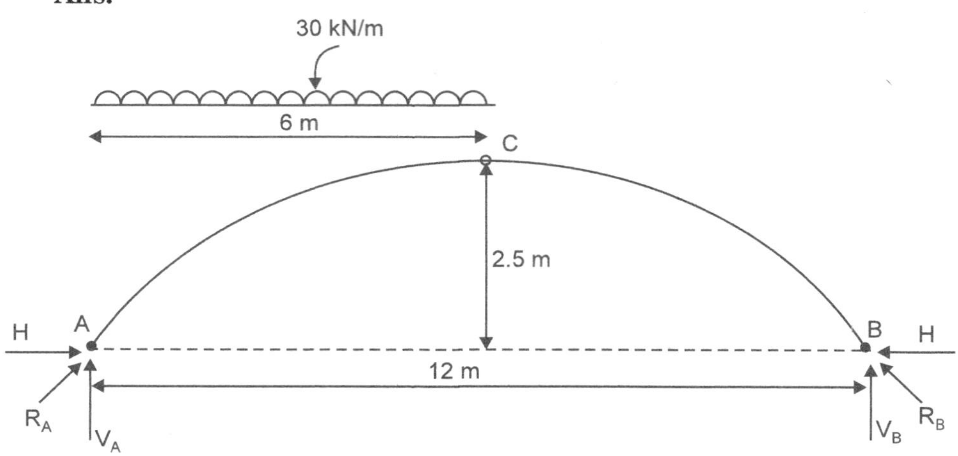

Fig. 5

Taking moments about A,

$$V_B \times 12 - 30 \times 6 \times 3 = 0$$

$$V_B = \frac{30 \times 6 \times 3}{12} = 45 \text{ kN}$$

$$V_A = 30 \times 6 - 45 = 135 \text{ kN}$$

Taking moments about C, $V_B \times 6 - H \times 2.5 = 0$

$$H = \frac{45 \times 6}{2.5}$$

$$H = 108 \text{ kN}$$

$$R_A = \sqrt{V_A^2 + H^2} = \sqrt{(135)^2 + (108)^2} = 172.88 \text{ kN}$$

$$R_B = \sqrt{V_B^2 + H^2} = \sqrt{(45)^2 + (108)^2} = 117 \text{ kN}$$

$\therefore$ $R_A = 172.88$ kN; $R_B = 117$ kN

12. Explain with the aid of a sketch, the normal thrust and radial shear in an arch rib.

Ans.

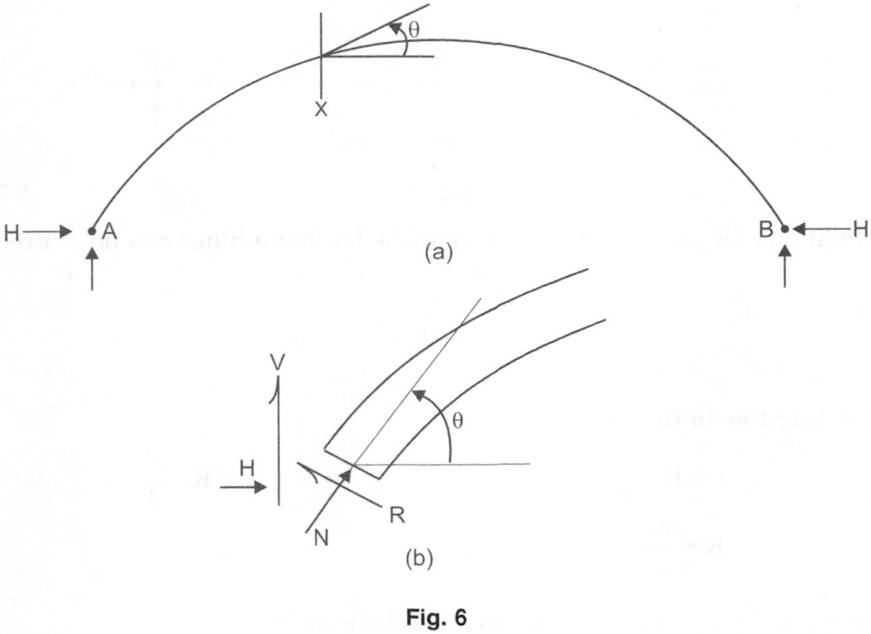

Fig. 6

Let us take a section X of an arch [Fig. 6 *(a)*]. Let θ be the inclination of the tangent at X. If H is the horizontal thrust and V the net vertical shear at X, from the free body of the RHS of the arch, it is clear that V and H will have normal and radial components given by,

$$N = H \cos \theta + V \sin \theta$$

$$R = V \cos \theta - H \sin \theta$$

13. A three hinged parabolic arch is loaded with uniformly distributed load throughout the span. What is the approximate variation of bending moment ?

Ans. When a three hinged parabolic arch (actual arch) is subjected to uniformly distributed load, the bending moment diagram (theoretical arch) will also be a parabola and coincident with the arch. Hence, the bending moment, according to Eddy's theorem, everywhere will be zero.

14. Which of the two arches, viz. circular and parabolic is preferable to carry a uniformly distributed load ? Why ?

Ans. Parabolic arches are preferable to carry distributed loads. Because both the shape of the arch and the shape of the bending moment diagram are parabolic. Hence the intercept between the theoretical arch and actual arch is zero everywhere. Hence, the bending moment at every section of the arch will be zero. The arch will be under pure compression which will be economical.

15. Obtain the equation to a parabolic arch (span 'l' and rise 'r') with origin at the left hinge.

Ans.

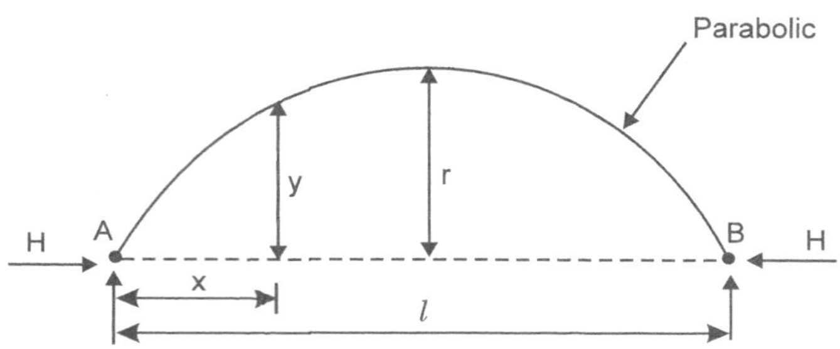

Fig. 7

The equation of a parabola, with origin at the left hand hinge can be written as,

$$Y = Kx\,(l - x) \qquad ...(1)$$

where K is a constant.

At $x = \frac{l}{2}$, $y = r$

Substituting this in the equation (1)

$$r = K \cdot \frac{l}{2}\left(l - \frac{l}{2}\right) \quad \Rightarrow \quad r = K \cdot \frac{l^2}{4}$$

$$K = \frac{4r}{l^2}$$

$\therefore \quad y = \frac{4r}{l^2}\,x\,(l - x)$ is the equation to a parabolic arch.

16. Give the equation for a parabolic arch whose springings are at different levels.

Ans. In the case of a parabolic arch with supports at different levels, the equation to the arch with the crown as the origin is

$$y = \left(\frac{4y_c}{l^2}\right) x^2$$

This will be the same on both sides of the arch [Fig. 7(*a*)]

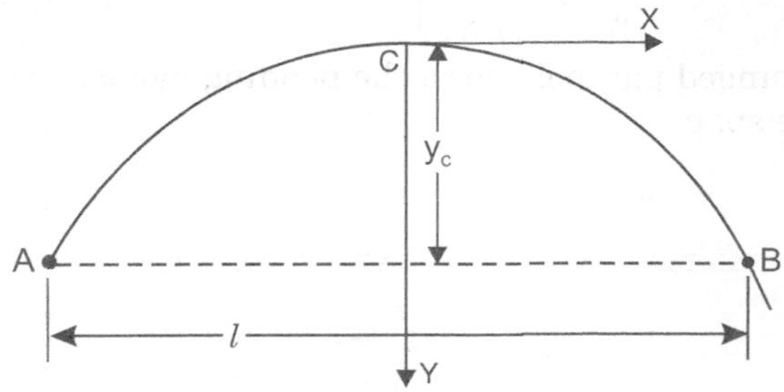

Fig. 7 (a)

17. What is the difference between the basic actions of an arch and a suspension cable ?

Ans. An arch is essentially a compression member which can also take bending moments and shears. Bending moment and shears will be absent if the arch is parabolic and the loading uniformly distributed.

A cable can take only tension. A suspension bridge will therefore have a cable and a stiffening girder. The girder will take the bending moment and shears in the bridge and the cable, only tension.

Because of the thrusts in cables and arches, the bending moments are considerably reduced.

If the load on the girder in uniform, the bridge will have only cable tension with no bending moment on the girder.

18. Find the level of hinge B from C of the parabolic arch shown in Fig. 7(*b*)

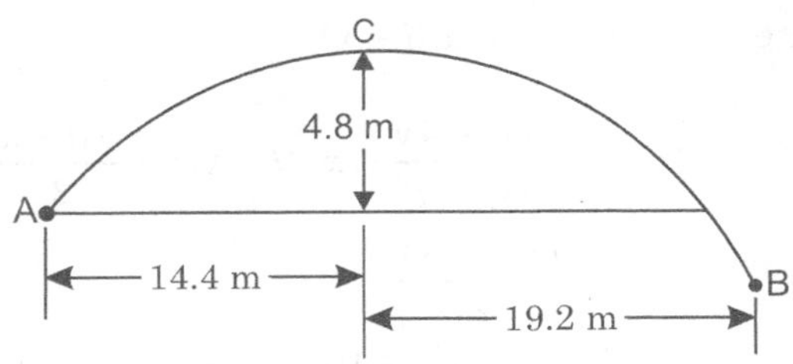

Fig. 7 (b)

Ans.

$$y = \frac{4y_c}{l^2} . x^2 ; l = 14.4 \times 2 = 28.8 \text{ m}$$

$$y_B = \frac{4y_c}{l^2} \times 19.2^2 = \frac{4 \times 4.8}{28.8^2} \times 19.2^2$$

$$y_B = 8.53 \text{ m from C}$$

Or

$$l_1 = 14.4 \text{ m} ; l_2 = 19.2 \text{ m}$$

$$y_1 = 4.8 \text{ m} ; y_2 = \text{required}$$

$$\frac{y_1}{y_2} = \left(\frac{l_1}{l_2}\right)^2$$

$$\frac{4.8}{y_2} = \left(\frac{14.4}{19.2}\right)^2 \qquad \therefore \quad y_2 = 8.53 \text{ m}$$

19. Prove that in a 3 hinged parabolic arch the bending moment is zero everywhere, under an udl all over the span.

Ans.

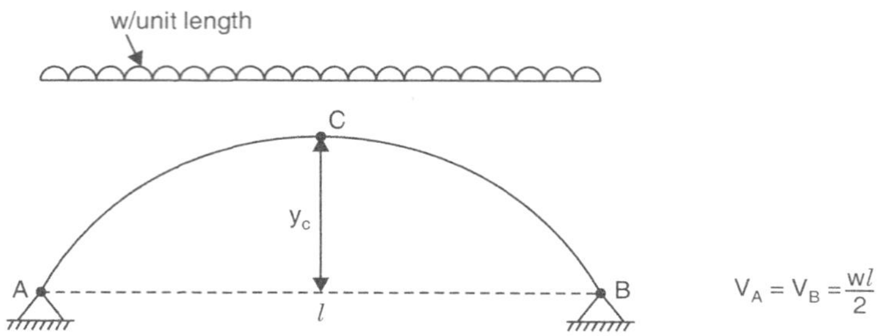

Fig. 8

Since BM at C is zero,

$$V_A \cdot \frac{l}{2} - w \cdot \frac{l}{2} \cdot \frac{l}{4} - H \cdot y_c = 0$$

$$H \cdot y_c = \frac{wl}{2} \cdot \frac{l}{2} - \frac{wl}{2} \cdot \frac{l}{4}$$

$$H = \frac{wl^2}{8y_c}$$

at any x, the ordinate $$y = \frac{4y_c}{l^2} x.(l - x)$$

$$H \cdot y = \frac{wl^2}{8y_c} \cdot \frac{4y_c}{l^2} \cdot x \cdot (l - x) = \frac{wx(l - x)}{2}$$

$$\mu_x = \frac{wx(l - x)}{2}$$

$$BM = \mu_x - Hy = 0.$$

20. A three hinged semicircular arch carries a point load of 100 kN at the crown. The radius of the arch is 4 m. Find the horizontal reactions at the supports.

Ans. $$V_A = V_B = 50 \text{ kN}$$

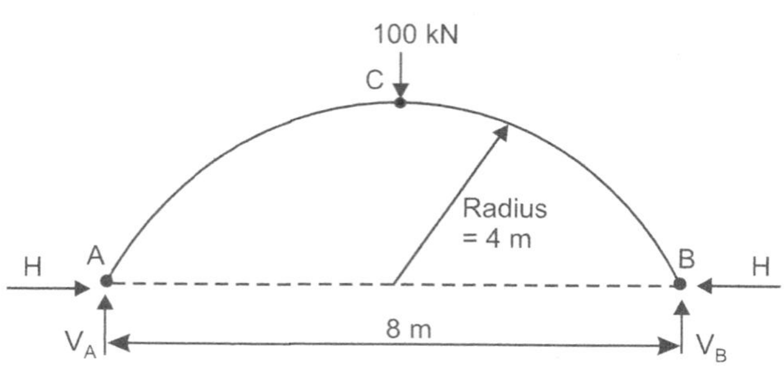

Fig. 8 (a)

Equating the moment about C to zero, $V_A \times 4 - H \times 4 = 0$

$$H = V_A$$

Horizontal reaction, H = 50 kN.

21. A three hinged semicircular arch of radius 10 m carries a udl of 2 kN/m over the whole span. Determine the horizontal and vertical reactions at the supports.

Ans.

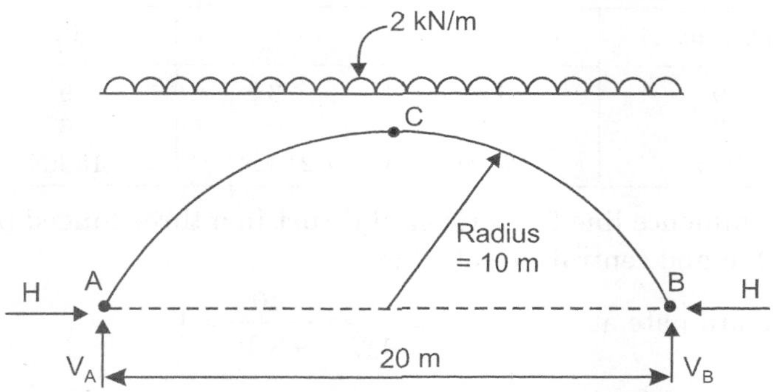

Fig. 9

$$V_A = V_B = \frac{\text{Total load}}{2} = \frac{2 \times 20}{2} = 20 \text{ kN}$$

Equating moments about the crown C to zero,

$$V_A \times 10 - 2 \times 10 \times \frac{10}{2} - H \times 10 = 0$$

$$20 \times 10 - 20 \times 5 = H \times 10$$

$$H = \frac{200 - 100}{10} = 10 \text{ kN}$$

$\therefore$ At the supports, vertical reactions = 20 kN

horizontal reactions = 10 kN

22. Determine H, V_A and V_B in the semicircular arch shown in Fig. 10.

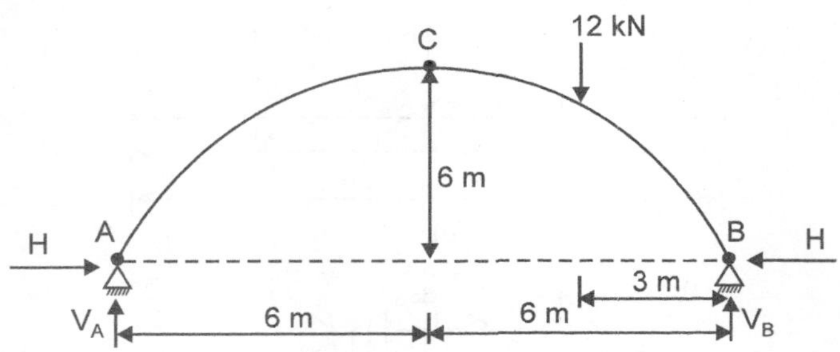

Fig. 10

Ans. Equating moments about A to zero,

$$V_B \times 12 - 12 \times 9 = 0; \quad V_B = 9 \text{ kN}$$

$$V_A = 3 \text{ kN}$$

Equating moments to the left of C to zero,

$$H = V_A = 3 \text{ kN}$$

23. Using the result of problem 22, find H, V_A and V_B when the arch has one more load of 24 kN at 3 m from A.

Ans. The results are tabulated as under:

Dist. From A	*Load*	V_A	V_B	*H*
9	12	3	9	3
3	24	18	6	6
9, 3	12, 24	21 kN	15 kN	9 kN

24. Draw the influence line for horizontal thrust in a three hinged parabolic arch having a span of 40 m and central rise of 10 m.

Ans. Max. ordinate at the centre $= \dfrac{l}{4y_c} = \dfrac{40}{4 \times 10} = 1$

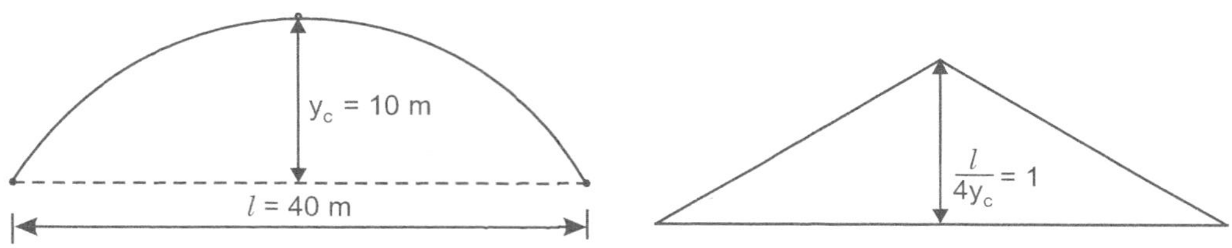

Fig. 11. ILD for horizontal thrust.

25. Draw the influence line diagram for bending moment at a section X at a distance x from the left end of a three hinged parabolic arch of span 'l' and rise 'h'.

Ans.

$$M_x = \mu_x - Hy$$

The max orinates arc, $\dfrac{x(l-x)}{l}$ and $\dfrac{l}{4h} \cdot \dfrac{4h}{l^2} \cdot x(l-x)$

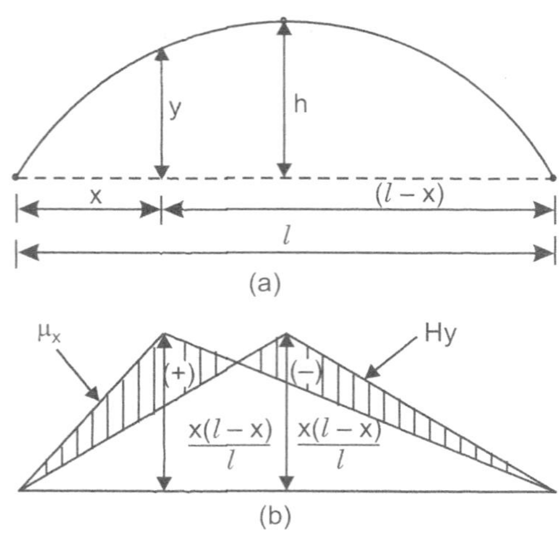

Fig. 12. ILD for bending moment at a section X.

26. Indicate the positions of a moving point load for maximum negative and positive bending moments in a three hinged arch.

Ans.

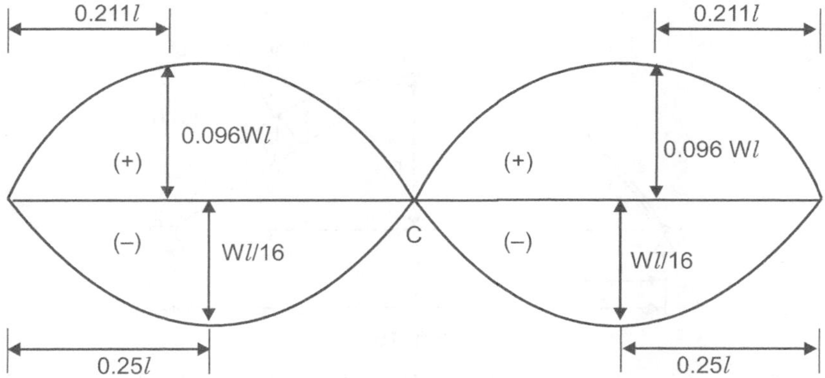

Fig. 13. Max. BMD due to point load.

Considering a three hinged parabolic arch of span 'l' and subjected to a moving point load W, the position of the point load for

(*i*) Maximum negative bending moment is 0.25l from end supports

(*ii*) Maximum positive bending moment is 0.211l from end supports.

27. Draw the influence line for radial shear at a section of a three hinged arch.

Ans. Radial shear is given by

$$F_x = V_x \cos\theta - H \sin\theta$$

where θ is the inclination of tangent at X.

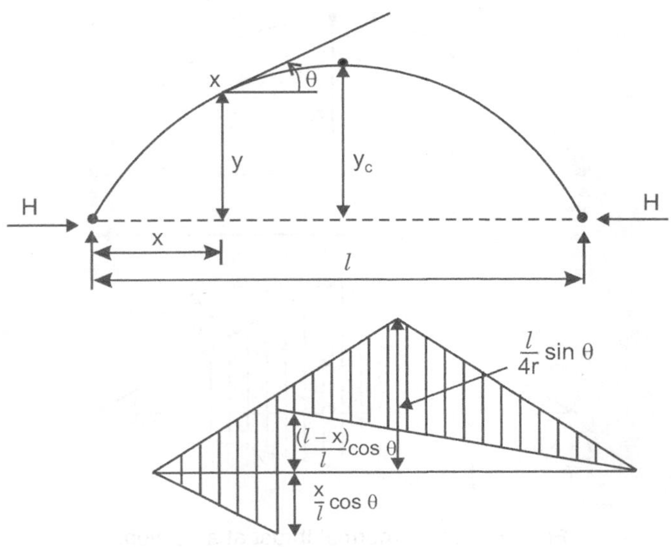

Fig. 14. ILD for radial shear at X.

28. An arch has hinges at supports and at crown. The span is 'l' and rise is 'r'. A concentrated load W acts at $l/3$ from left support. Sketch the line of thrust diagram.

Ans.

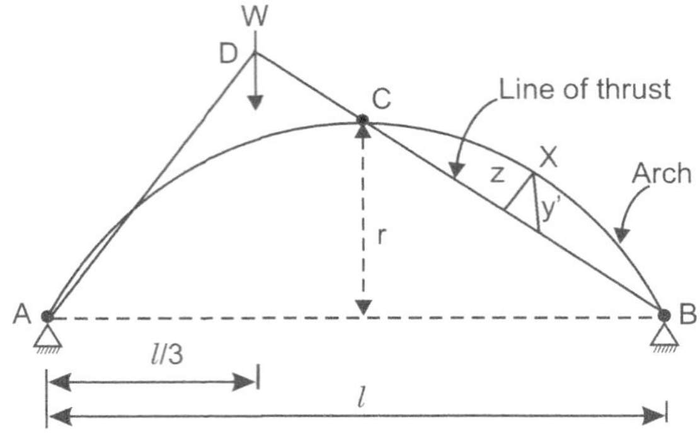

Fig. 15

The line of thrust is actually the funicular polygon for the load and reactions. It has to pass through the hinge at the crown.

DA and DB are the lines along which the load at D will get transmitted to the supports A and B.

If the arch were along DA and DB there would have been no bending moment. As such at any section X, the thrust $T.z$ will be the bending moment at X. This is the same as the horizontal component H of T, multiplied by y'. ($T.z = H.y'$)

29. Sketch the influence line diagram for the normal thrust at a section X of a symmetric three hinged parabolic arch.

Ans. Normal thrust at X is given by $P = H \cos\theta + V_x \sin\theta$, where θ is the inclination of tangent at X (see Fig. 16).

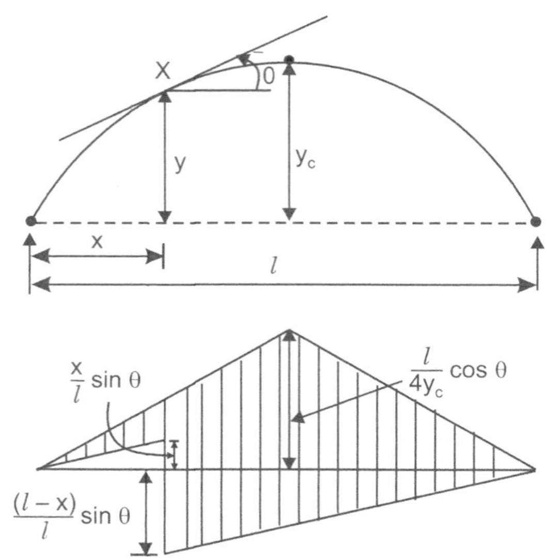

Fig. 16. ILD for normal thrust at a section.

30. What is reaction locus in a two hinged arch ?

Ans. Let us consider a moving concentrated load W on a 2 hinged arch. Let x be the variable distance of W from the left support. W gives rise to resultant reactions R_A and R_B at supports A and B.

Since there are only 3 forces, W, R_A and R_B on the system they must intersect at a point, say P, (at x from A).

The locus of P as x varies from zero to full span is the reaction locus.

It is the locus of the point of intersection of the 2 reactions R_A and R_B. See Fig. 16 (*a*).

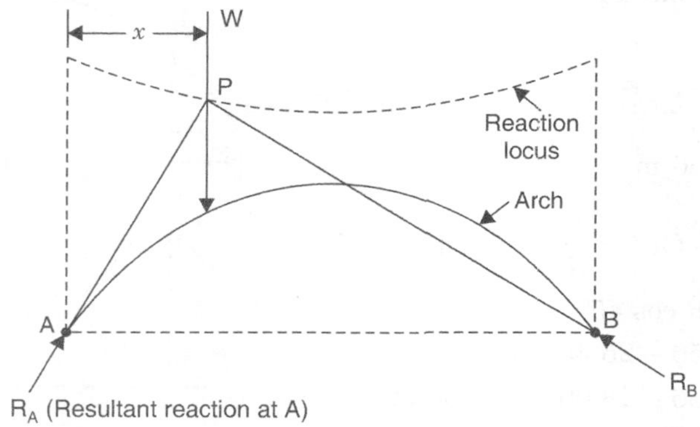

Fig. 16 (a). Reaction Locus.

31. Fig. 16 (*b*) shows the reaction locus of a 2 hinged arch. Using the locus, find (*i*) the lines of action of reactions, (*ii*) the linear arch, and (*iii*) the regions of positive and negative bending moments, when a load W is at 8 m from A.

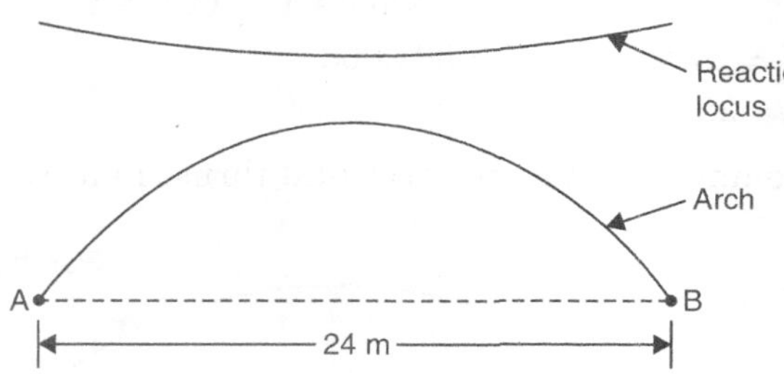

Fig. 16 (b)

Ans.

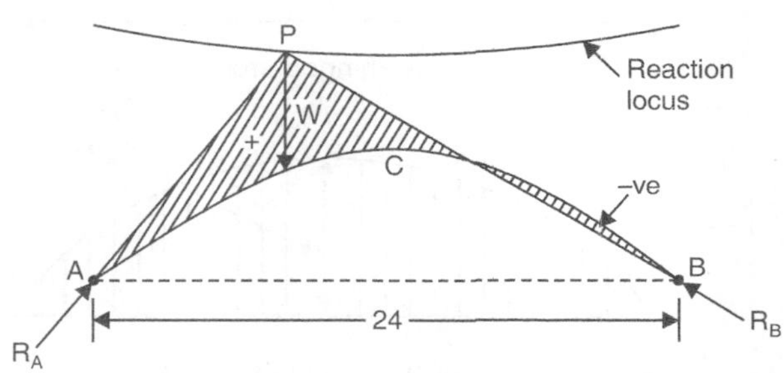

Fig. 16 (c)

Produce the line of action of W to cut reaction locus at P. AP and BP are the lines of R_A and R_B.

APB is the linear arch. The space between the linear arch APB and Real arch ACB is the BM diagram.

32. A symmetrical two hinged segmental arch of span 36 m subtends an angle of $2\phi = 90°$ at the centre. Obtain the radius and rise of the arch.

Ans.

$$\frac{AC'}{OA} = \frac{18}{R} = \sin 45°$$

$$\frac{18}{\sin 45°} = R$$

$$R = 25.456 \text{ m}$$

$$y_c = OC - OC' \qquad \frac{OC'}{R} = \cos 45°$$

$$= R - R\cos 45°$$

$$= 25.456 - 25.456 \cos 45°$$

$$= 25.456 - 18.000 = 7.456 \text{ m}$$

$$\therefore \quad R = 25.456 \text{ m}; \; y_c = 7.456 \text{ m}$$

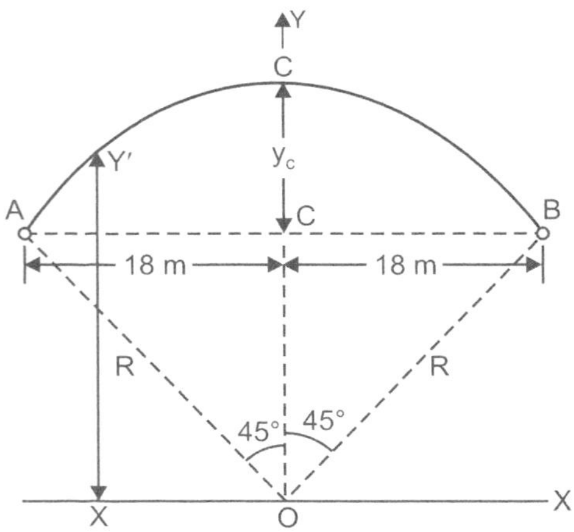

Fig. 17

33. In the above arch, obtain the rise of the arch at 6 m from the left hand support.

Ans. X = 18 – 6 = 12 m from the centre of the span

The circle can be given by

$$X'^2 + Y'^2 = R, \qquad \text{where } Y' = OC' + Y$$

at $\quad X' + 12, \; Y'^2 = 25.456^2 - 12^2 = 504.008$

$$Y' = 22.45; \; Y = 4.45 \text{ m}$$

34. Sketch the influence line diagram for horizontal thrust in a two hinged parabolic arch.

Ans.

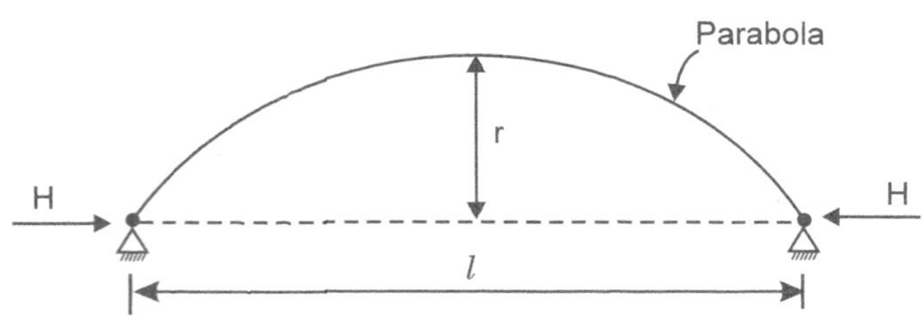

Two hinged arch

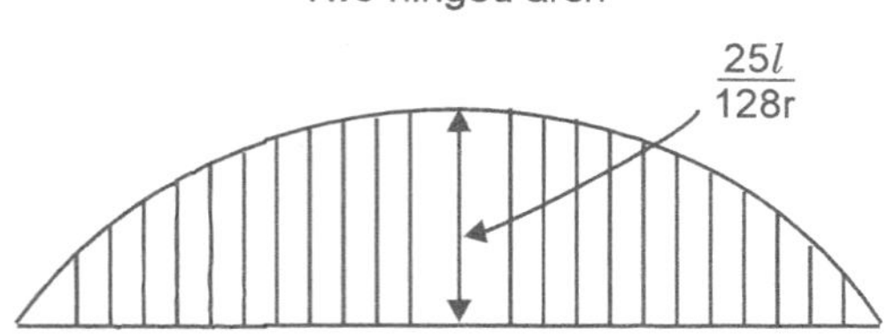

Fig. 18. ILD for horizontal thrust, H.

35. Distinguish between two hinged and three hinged arches.

Ans.

S.No.	*Two hinged arches*	*Three hinged arches*
1	Statically indeterminate to first degree.	Statically determinate.
2	Might develop temperature stresses.	Increase in temperature causes increase in central rise. No stresses.
3	Structurally more efficient.	Easy to analyse. But, in construction, the central hinge may involve additional expenditure.
4	Will develop stresses due to sinking of supports.	Since this is determinate, no stresses due to support sinking.

36. Explain rib-shortening in the ease of arches.

Ans. In a 2 hinged arch, the normal thrust which is a compressive force along the axis of the arch will shorten the rib of the arch. This is turn will release part of the horizontal thrust.

Normally, this effect is not considered in the analysis (in the case of two hinged arches).

Depending upon the importance of the work we can either take into account or omit the effect of rib-shortening. This will be done by considering (or omitting) strain energy due to axial compression along with the strain energy due to bending in evaluating H.

37. State the effects of temperature on three hinged and two hinged arches.

Ans. When subjected to temperature change:

1. A three hinged arch will simply change its central rise if there is expansion or contraction due to change in temperature.

2. A two hinged arch will develop internal stresses due to changes in temperature. This will develop a horizontal thrust and consequent changes in bending moment, shears, etc.

38. Give the expression for the temperature effect on a two hinged arch.

Ans. Additional horizontal thrust due to temperature change,

$$H_1 = \frac{l\,\alpha\,TEI}{\int y^2 ds}$$

with usual notations.

39. What are the methods used for the analysis of fixed arches?

Ans. The methods used for the analysis of fixed arches are :

1. Using Castigliano's theorem.
2. Elastic centre method.
3. Column analogy method.

40. Explain the effect of yielding of support in the case of an arch.

Ans. Yielding of supports has no effect in the case of a 3 hinged arch which is determinate. These displacements must be taken into account when we analyse 2 hinged or fixed arches as under

$$\frac{\partial U}{\partial H} = \Delta H \text{ instead of zero}$$

$$\frac{\partial U}{\partial V_A} = \Delta V_A \text{ instead of zero}$$

Here U is the strain energy of the arch and ΔH and ΔV_A are the displacements due to yielding of supports.

41. For the arch shown in Fig. 19, find the normal thrust at a section 5 m from the left end, given θ at D = 20°33′.

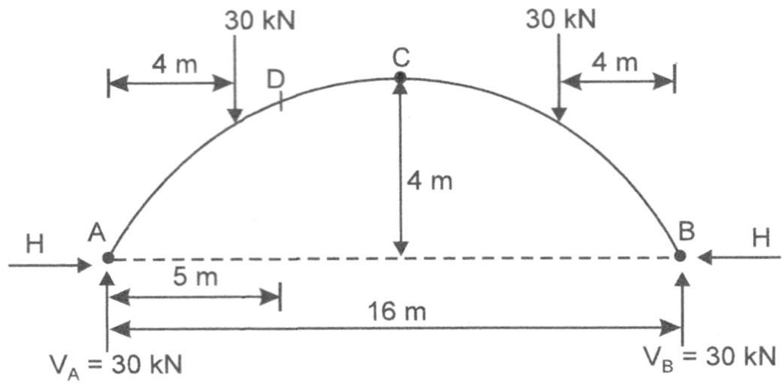

Fig. 19

Ans. By symmetry $V_A = V_B = 30$ kN

Net vertical shear at D = 30 – 30 = 0

Equating moments about C to zero, $V_A \times 8 - 30 \times 4 - H \times 4 = 0$

$$30 \times 8 - 30 \times 4 - H \times 4 = 0;\ H = 30 \text{ kN}$$

Normal thrust at D, $P = V \sin\theta + H \cos\theta$

$$= 0 + 30 \cos 20°33' = 28.09 \text{ kN}$$

$\therefore$ $P = 28.09$ kN

42. A three hinged parabolic arch has a horizontal span of 36 m with a central rise of 6 m. A point load of 40 kN moves across the span from the left to the right. What is the absolute maximum positive bending moment that will occur in the arch ?

Ans.

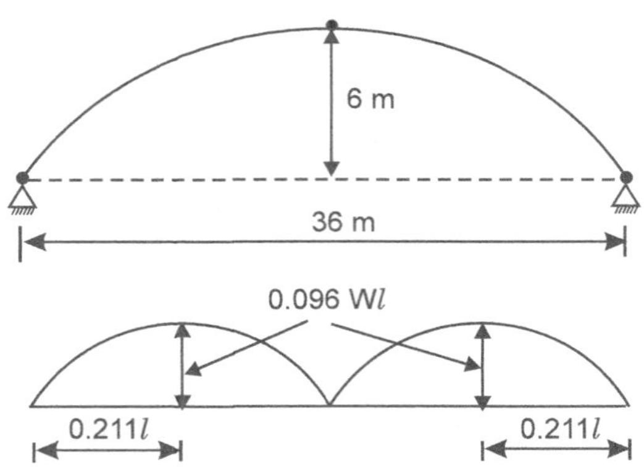

Fig. 20. Max. +ve BMD due to single concentrated moving load W.

For a single concentrated load moving from one end to the other,

Absolute maximum positive bending moment

$$= 0.096\, Wl = 0.096 \times 40 \times 36 = 138.24 \text{ kNm}$$

This occurs at $0.211l = 0.211 \times 36 = 7.596$ m from the ends.

$\therefore$ Absolute maximum positive bending moment = 138.24 kNm at 7.596 m from the ends.

43. A three hinged parabolic arch of span 30 m has a central rise of 5 m. Find the increase in the rise of the arch if the temperature rises through 30°C. Take coefficient of linear expansion of the arch material as 12×10^{-6} /°C.

Ans. $l = 30$ m ; $y_c = 5$ m ; $\Delta y_c = ?$; T = 30°C ; $\alpha = 12 \times 10^{-6}$ / °C

Increase in the rise of the arch

$$\Delta y_c = \frac{l^2 + 4y_c^2}{4y_c}\alpha T = \frac{30^2 + 4 \times 5^2}{4 \times 5} \times 30 \times 12 \times 10^{-6} = 0.018 \text{ m} = 18 \text{ mm}$$

$$\Delta y_c = 18 \text{ mm.}$$

44. Determine the support reactions in the arch shown in Fig. 21.

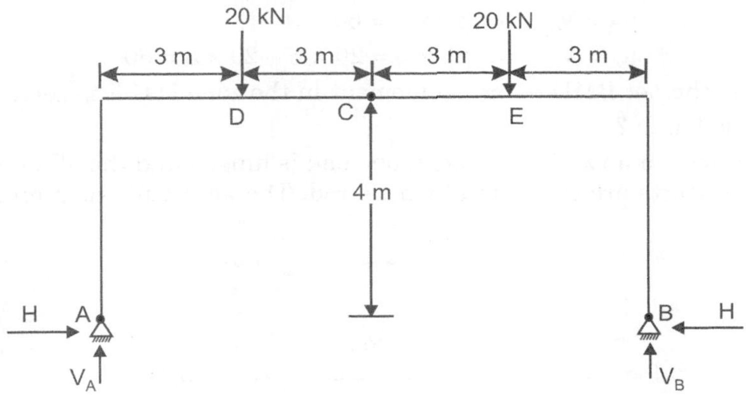

Fig. 21

Ans. $V_A = V_B = 20$ kN

Equating moment at C to zero,

$$V_A \times 6 - 20 \times 3 - H \times 4 = 0$$

$$H = \frac{20 \times 6 - 20 \times 3}{4} = 15 \text{ kN}$$

45. Sketch the BMD for the arch shown in Fig. 21.

Ans. The BMD [Fig. 22 (*c*)] is obtained by superposing the μ diagram [Fig. 22 (*a*)] and the Hy diagram [Fig. 22 (*b*)].

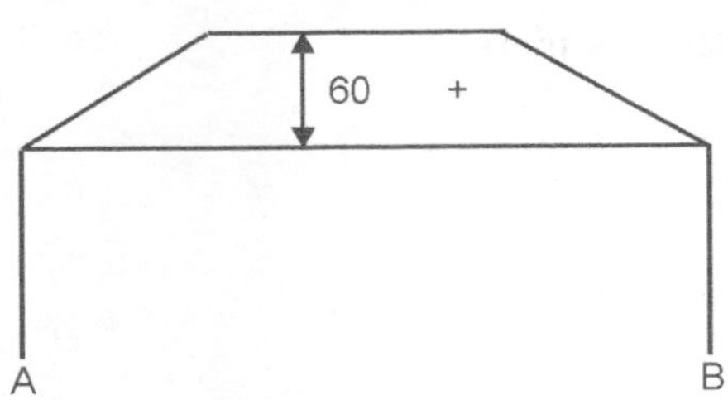

Fig. 22. (a) μ Diagram.

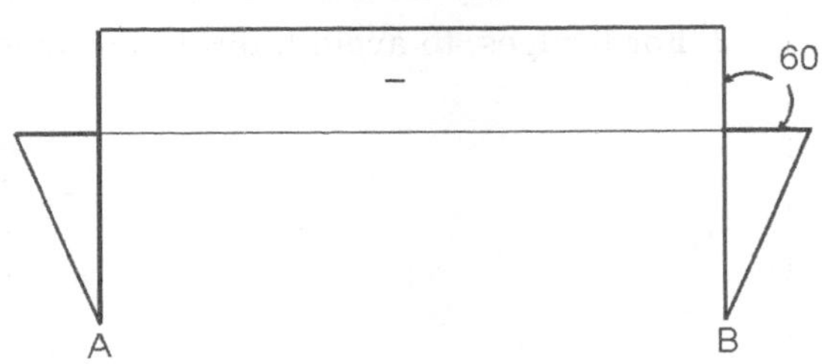

Fig. 22. (b) Hy Diagram.

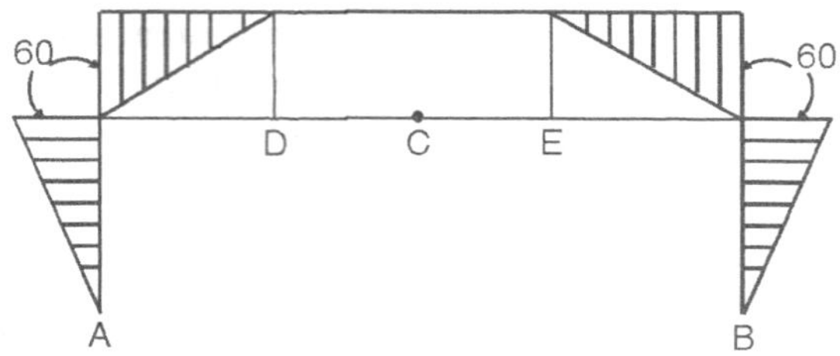

Fig. 22. (c) Net BMD.

$$H\,y_c = -15 \times 4 = -60 \text{ kNm}$$
$$\mu_D = V_A \times 3 = 20 \times 3 = 60$$
$$\mu_C = V_A \times 6 - 20 \times 3 = 20 \times 6 - 20 \times 3 = 60$$

Fig. 22 (*c*) is the net BMD. Bending moment in the zone D C E is zero.

46. What is a tied arch ?

Ans. Sometimes in a two hinged arch, one end is hinged and the other end is placed on rollers. The two supports are connected by a tie rod. The arch with such an arrangement is called a tied arch.

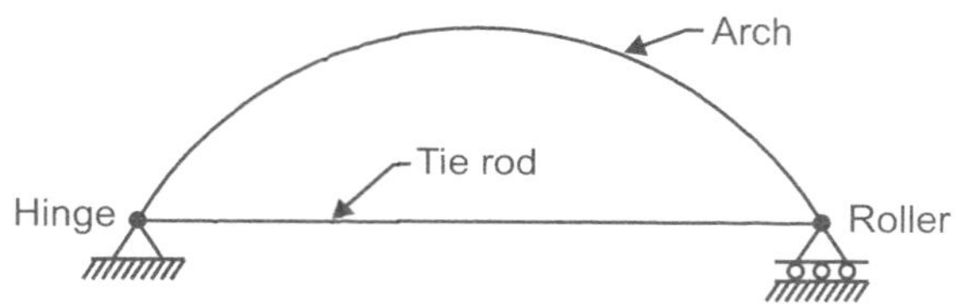

Fig. 23. Tied Arch.

47. What will be the force acting in the tie rod of a two hinged tied arch ?

Ans. The roller end of a loaded tied arch tends to move which will be restricted by the presence of the tie rod. Hence the movement of the roller end and the extension of the tie rod will be the same. Hence,

Tensile force in the tie rod = Horizontal thrust in the arch.

48. Mention two applications of tied arches.

Ans. Tied arches can be used:

1. Economically for a roof resting on walls incapable of resisting thrust.
2. For bridges, to avoid lateral movement of abutments piers.

14

Cables and Suspension Bridges

1. What are cable structures ? Give examples.

Ans. Structures in which the cables act as the main load bearing members are called cable structures. Cables will be in the form of wire ropes or parallel wires joined with clips. Cables are flexible, will take only direct tension and cannot take any bending moment.

***Example*:** Suspension bridges, cable stayed bridges, Rope car, winch car.

2. What are the similarities in the structural actions of arches and suspension bridges?

Ans. Arches develop compression in addition to bending moments. The compression part is contributed by the horizontal thrust H. H is responsible for the drastic reduction in bending moment. The arch rib is designed to take both compression and bending moment.

In a cable-stiffening girder combination (suspension bridge), the reduction in bending moment in the girder is brought about by the cable tension. Here the girder takes the bending moment and the cable takes the tension.

3. A suspension cable of span 150 m and dip 15 m is supported at the same level and carries a load of 6 kN/m run of horizontal span. Find the maximum tension in the cable.

Ans.

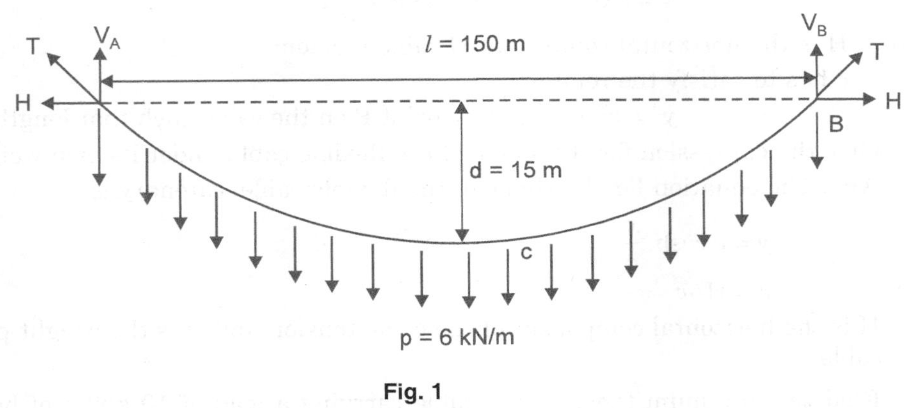

Fig. 1

$$V_A = V_B = \frac{pl}{2} = \frac{6 \times 150}{2} = 450 \text{ kN}$$

$$H = \frac{pl^2}{8d} = \frac{6 \times (150)^2}{8 \times 15} = 1125 \text{ kN}$$

$$T_{max} = \sqrt{V^2 + H^2} = \sqrt{(450)^2 + (1125)^2} = 1211.66 \text{ kN}$$

∴ Maximum tension in the cable, T_{max} = 1211.66 kN.

4. The horizontal span of a parabolic suspension cable is 100 m and the central dip is 8 m. Find the length of the cable.

Ans. l = 100 m ; d = 8 m

Length of the cable, $$S = l + \frac{8}{3}\frac{d^2}{l} = 100 + \frac{8}{3} \times \frac{8^2}{100} = 101.707 \text{ m}$$

∴ Length of the cable, S = 101.707 m.

5. A cable is suspended over a span l. There is no load on the cable except its own weight. Write down the equation to the cable indicating clearly the co-ordinate axes.

Ans.

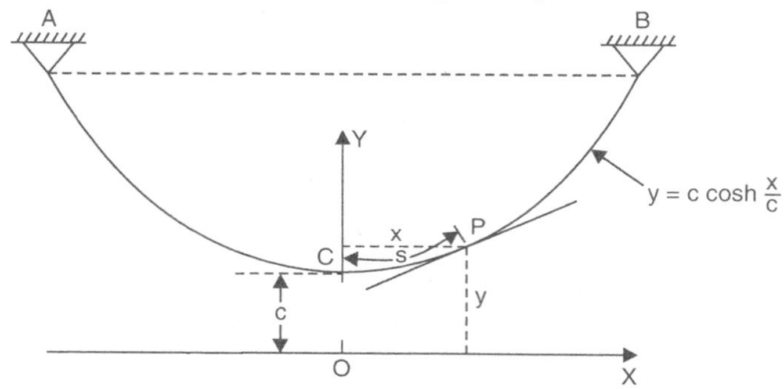

Fig. 1 (a)

The equation to the cable suspended under its own weight w per unit length is

$$y = c \cosh \frac{x}{c}, \quad \text{where } c = \frac{H}{w},$$

where H is the horizontal component of cable tension

c has to satisfy the relation

$y^2 = s^2 + c^2$ for any point P on the cable such that length CP = s.

6. Give the expression for the shape of the flexible cable under its own weight.

Ans. The equation for the shape of the flexible cable, catenary is

$$y = c \cosh \frac{x}{c}$$

where $c = H/w$

H is the horizontal component of the cable tension and w is the weight per unit length of the cable.

7. Find the minimum tension in a cable carrying a load of 10 kN/m of horizontal span of 80 m. The supports are at the same level and the central dip is 4 m.

Ans. l = 80 m ; d = 8 m ; p = 10 kN/m ; T_{min} = ?

The minimum tension will occur at the lowest point in the cable and its value will be equal to H.

Since the cable is supported at the same level,

$$T_{min} = H = \frac{pl^2}{8d} = \frac{10 \times 80^2}{8 \times 8} = 1000 \text{ kN}.$$

8. Write down an expression to find the length of a cable.

Ans. Length of a suspension cable :

(*i*) Supports at the same level :

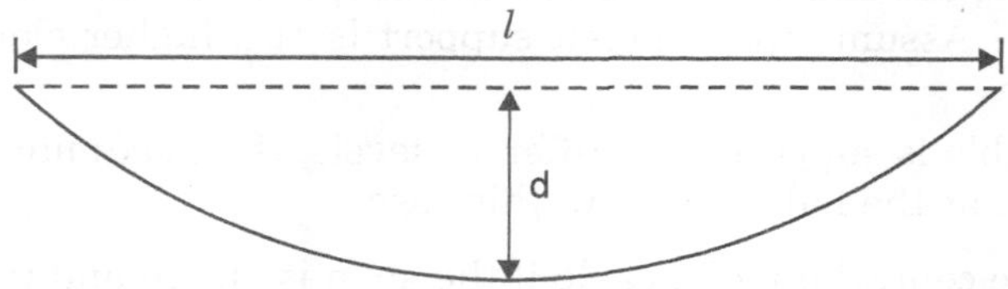

Fig. 2 (a)

Length of the cable, $$S = l + \frac{8}{3}\frac{d^2}{l}$$

(*ii*) Supports at the different levels :

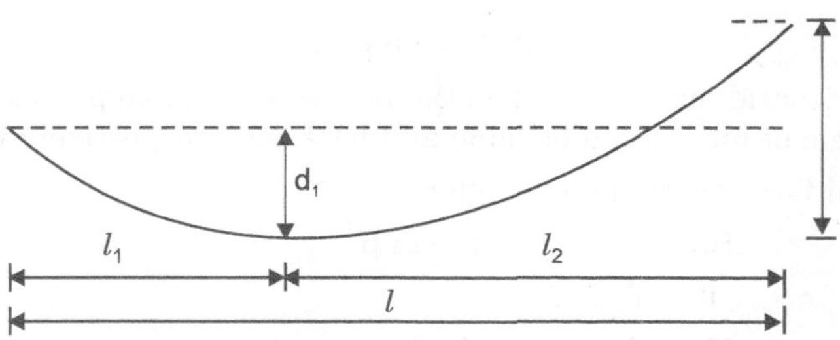

Fig. 2 (b)

Length of the cable, $$S = l + \frac{2}{3}\frac{d_1^2}{l_1} + \frac{2}{3}\frac{d_2^2}{l_2}$$

9. A cable of span '*l*' and central dip '*d*' is subjected to a uniformly distributed load w per unit horizontal length. What is the horizontal component of tension in the cable ?

Ans. Span = l ; Central dip = d ; udl = w/unit horizontal length.

$\therefore$ Horizontal component of tension, $H = \dfrac{wl^2}{8d}$

10. A cable of span 100 m and central dip 16 m carries a udl of 16 kN/m all along its span. Find the cable tension.

Ans. $$V_A = V_B = \frac{1}{2} \times 100 \times 16 = 800 \text{ kN}$$

$$H = \frac{wl^2}{8d} = \frac{16 \times 100 \times 100}{8 \times 16} = 1250 \text{ kN}$$

$$T = \sqrt{V^2 + H^2} = \sqrt{(800)^2 + (1250)^2} = 1484.08 \text{ kN}$$

11. What is true shape of suspension cable which is supported between two pegs ? The cable supports only its own weight.

Ans. Catenary.

12. Where do you get the maximum tension in a suspension cable, when it is supported at different levels ? Assume that the left support is at a higher elevation than the right support.

Ans. When a cable is supported at different levels, the maximum tension will occur at the higher support *i.e.*, at the left support in this case.

13. What will be the central dip of a cable if the span is 100 m and the length is 102 m ?

Ans. Length of the cable, $L = l + \frac{8}{3}\frac{d^2}{l}$

$$102 = 100 + \frac{8}{3} \times \frac{d^2}{100}$$

$$d^2 = 2 \times \frac{3}{8} \times 100 = 75$$

$$d = \sqrt{75} = 8.66 \text{ m.}$$

14. What is the horizontal force exerted on the pier when a cable passes over a pulley ? Let α and β the angle of inclination of cable and back stay respectively with the pier.

Ans. Horizontal force exerted on the pier,

$$H_T = T_s \sin\alpha - T_A \sin\beta$$

$$T_s = T_A = T$$

$$\therefore \quad H_T = T(\sin\alpha - \sin\beta)$$

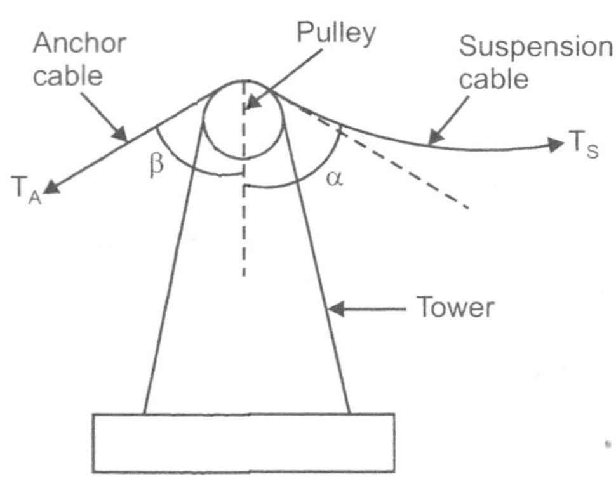

Fig. 3

15. What are the forces to be considered for the design of a tower with saddle piece ?

Ans. Since the saddle is placed on rollers, horizontal component of tension in the suspension cable = horizontal component of tension in the anchor cable.

i.e., $T_s \sin \theta_s = T_A \sin \theta_A = H$

Hence there will be no bending moment in the support tower.

Vertical force on the supporting tower,

$$V_T = T_s \cos \theta_s + T_A \cos \theta_A$$

Hence the supporting tower will be designed only for the vertical force, V_T. However, this force can act at an eccentricity, since the saddle can slide horizontally to accommodate the changes in θ_A and θ_s.

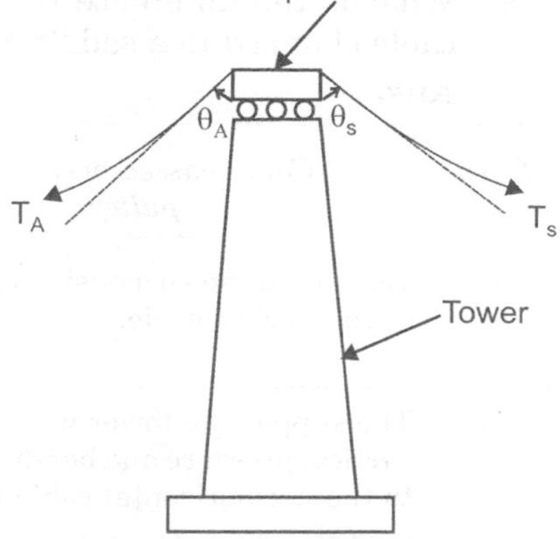

Fig. 4

16. What decides the height of pylons in a suspension bridge ?

Ans. The central dip of the cable determines the height of pylons.

The horizontal tension, H in a cable varies inversely as the central dip, d . $\left(H = \dfrac{wl^2}{8d} \right)$.

In addition to the dip, we have to consider the depth of the stiffening girder and the ground clearance of the deck (Fig. 5).

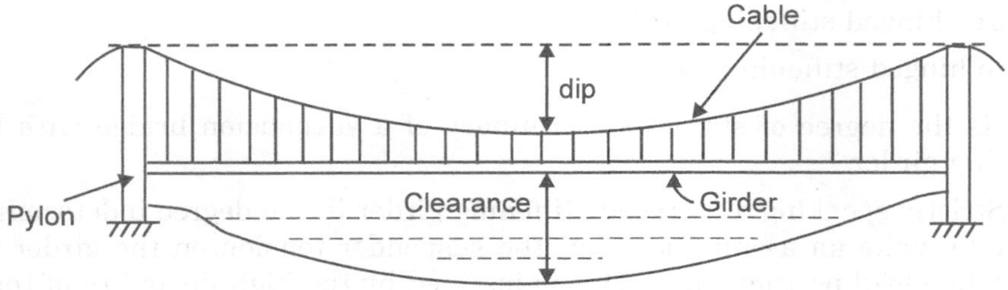

Fig. 5

17. What is a catenary ?

Ans. The shape that is taken up by a cable or rope freely suspended between two supports and under its own weight is called a catenary.

18. What is the relation between T and T_A of a cable clamped to a saddle carried on smooth rollers on the top of a pylon (Fig. 6)? Determine also the vertical reaction transferred to the pylon.

Ans.

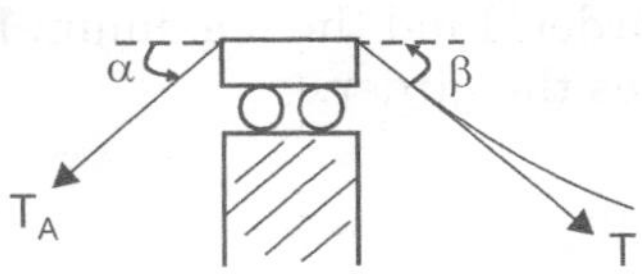

Fig. 6

Since the saddle is placed on rollers,

$$T \cos \beta = T_1 \cos \alpha$$

Vertical reaction transferred to the pylon

$$= T \sin \beta + T_1 \sin \alpha$$

19. What is the difference between the cable passed over a guide pulley at support and cable clamped to a saddle carried on smooth rollers on the top of pylon ?

Ans.

S.No.	*Cable passed over guide pulley*	*Cable passed over saddle placed on rollers*
1.	Tension in the suspension cable = Tension in the anchor cable.	Horizontal component of tension in the suspension cable = Horizontal component of tension in the anchor cable.
2.	The supporting tower will be subjected to vertical pressure and bending moment due to the net horizontal cable tension.	The supporting tower will be subjected to only vertical pressure due to the cable tension.

20. Why are stiffening girders necessary in suspension bridges ?

Ans. In suspension bridges, cables are the main load bearing members. Cables are flexible and change their shape depending upon the nature and position of the moving load on the deck slab. When stiffening girders, one on each side of the width of the bridge, are provided, they help the cable retain the parabolic shape during the passage of load. Also stiffening girders enable the suspension bridge decks to remain fairly level.

21. What are the types of stiffening girders ?

Ans. Stiffening girders are of two types:

1. Three hinged stiffening girders
2. Two hinged stiffening girders.

22. What is the degree of static indeterminacy of a suspension bridge with two hinged stiffening girder ?

Ans. Strictly speaking a 2 hinged stiffening girder is one degree indeterminate. But it is customary to make an assumption that the suspender tension on the girder is uniform, whatever the live load position on it. This is justified by the high flexibility of the cable and renders the structure determinate.

23. What are the differences between suspension bridges with two hinged and three hinged stiffening girders ?

Ans. Strictly speaking, a two hinged stiffening girder is statically indeterminate. But, we make a (justifiable) simplifying assumption that whatever be the loading on the deck, the cable receives a udl and keeps its parabolic shape.

So, for 2 hinged and 3 hinged girder bridges the shapes of the ild for H and w (loading on cable) are the same. Only the magnitudes differ.

24. A 100 m span cable has a 3 hinged stiffening girder. Find the maximum BM at P, 30 m from the left, when a point load of 10 kN crosses the girder.

Ans. Fig. 5 (*a*) is the ILD for BM at P

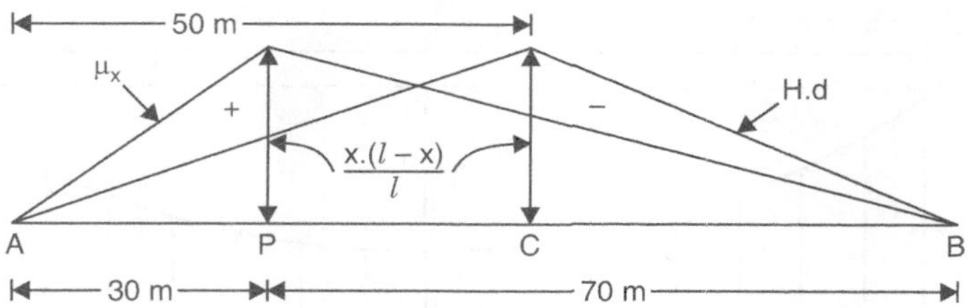

Fig. 6 (a). ILD for BM at P.

Max. positive ILO occurs at P

$$\text{ILO} = \frac{30 \times 70}{100}\left(1 - \frac{30}{50}\right) = 8.4$$

$$\text{Max BM} = 8.4 \text{ W}, \qquad \text{where W is the load}$$
$$= 8.4 \times 10 = 84 \text{ kNm.}$$

25. Draw the influence line for horizontal reaction on the cable in a three hinged stiffening girder bridge.

Ans.

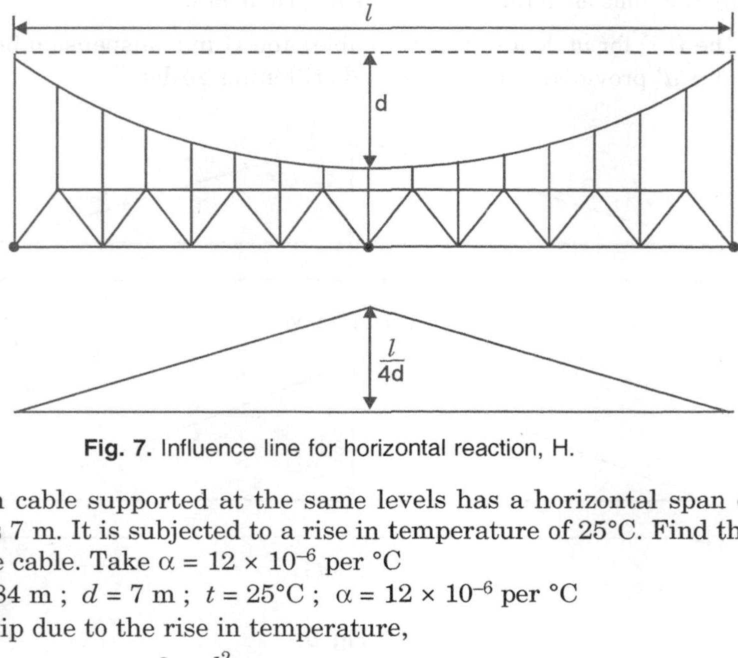

Fig. 7. Influence line for horizontal reaction, H.

26. A suspension cable supported at the same levels has a horizontal span of 84 m. The central dip is 7 m. It is subjected to a rise in temperature of 25°C. Find the increase in the dip of the cable. Take $\alpha = 12 \times 10^{-6}$ per °C

Ans. $l = 84$ m ; $d = 7$ m ; $t = 25$°C ; $\alpha = 12 \times 10^{-6}$ per °C

Increase in dip due to the rise in temperature,

$$\Delta d = \frac{3}{16} \times \frac{l^2}{d} \alpha\, t$$
$$= \frac{3}{16} \times \frac{(84)^2}{7} \times 12 \times 10^{-6} \times 25$$
$$= 0.0567 \text{ m} = 56.6 \text{ mm}$$

$\therefore$ Increase in dip = 56.7 mm.

27. Sketch the ILD for the load, w transmitted to a cable through a three hinged stiffening girder.

Ans.

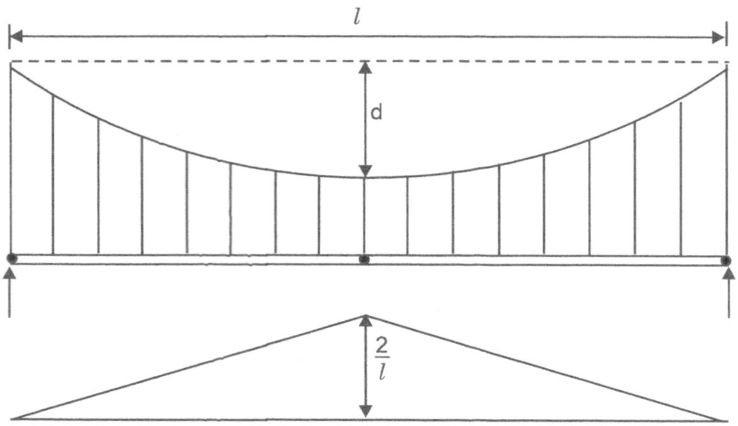

Fig. 8. ILD for w.

Note. For cables supported at the same level and stiffened by three hinged stiffening girders the ILD for bending moment at a section, max. BMD due to moving concentrated load W and max. BMD due to moving udl are the same as in the case of three hinged arches.

28. Sketch the ILD for w (loading on the cable) and H in a suspension bridge of span 'l' and central dip 'd' provided with a 2 hinged stiffening girder.

Ans.

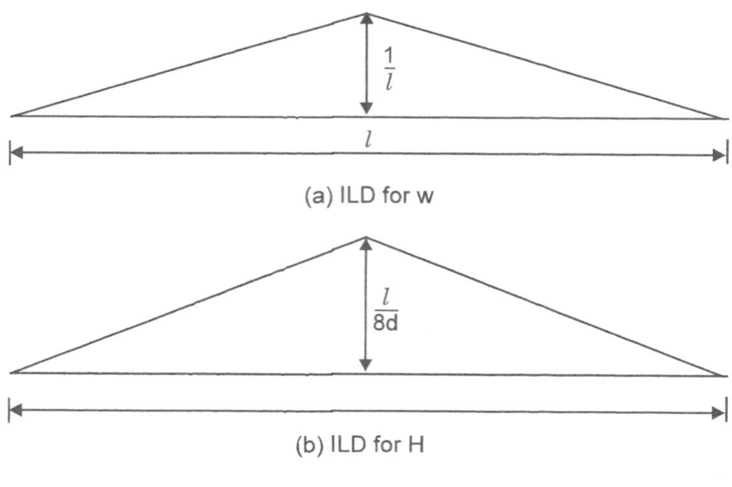

Fig. 9

29. A suspension cable having supports at the same level has a span of 45 m and a maximum central dip of 3.6 m. If the maximum tension in the cable is limited to 600 kN, what is the maximum udl that the cable can support ?

Ans. l = 45 m ; d = 3.6 m ; T_{max} = 600 kN ; w = ?

$$T_{max} = \sqrt{V^2 + H^2} = \sqrt{\left(\frac{pl}{2}\right)^2 + \left(\frac{pl^2}{8d}\right)^2}$$

Squaring both sides, $$T_{max}^2 = \left(\frac{pl}{2}\right)^2 + \left(\frac{pl^2}{8d}\right)^2$$

$$(600)^2 = \left(\frac{p \times 45}{2}\right)^2 + \left(\frac{p \times 45}{8 \times 3.6}\right)^2$$

$$(600)^2 = \left[\left(\frac{45}{2}\right)^2 + \left(\frac{45^2}{8 \times 3.6}\right)^2\right] p^2$$

$$p = \left[\frac{(600)^2}{\left(\frac{45}{2}\right)^2 + \left(\frac{45^2}{8 \times 3.6}\right)^2}\right]^{1/2} = \left(\frac{36 \times 10^4}{506.25 + 4943.85}\right)^{1/2}$$

$$= 8.127 \text{ kN/m}$$

udl that can be supported by the cable, w = 8.127 kN/m

30. A cable of horizontal span 50 m is to be used to support a concentrated of 30 kN at 15 m from the left support. The horizontal tension at the ends is 40 kN. Find the length of the suspension cable.

Ans.

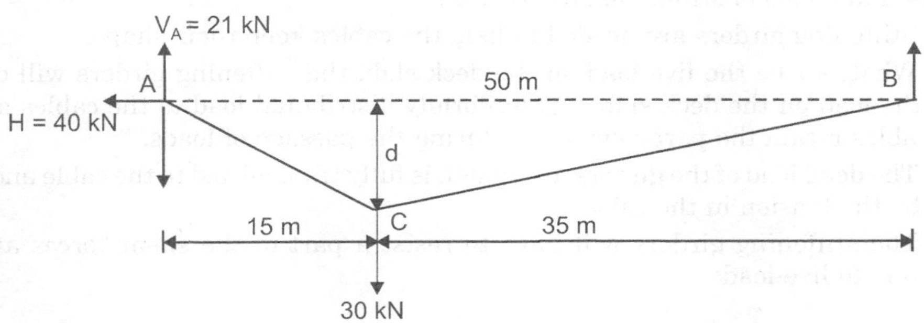

Fig. 10

$$V_A \times 50 - 30 \times 35 = 0$$

$$V_A = \frac{30 \times 35}{50} = 21 \text{ kN}; \quad V_B = 9 \text{ kN}$$

$$H = 40 \text{ kN (given)}$$

Taking moments about C,

$$V_A \times 15 - H \times d = 0$$
$$21 \times 15 - 40 \times d = 0$$

$$d = \frac{21 \times 15}{40} = 7.875 \text{ m}$$

$$AC = \sqrt{(15)^2 + (7.875)^2} = 16.941 \text{ m}$$

$$CB = \sqrt{(35)^2 + (7.875)^2} = 35.875 \text{ m}$$

∴ Length of the cable = 16.941 + 35.875 = 52.816 m.

31. A suspension cable of span 110 m and central dip 10 m is subjected to a rise in temperature of 28°C. Coefficient of linear expansion of the material is 12×10^{-6} per °C. Find the change in the horizontal tension, if the cable carries a uniform load of 30 kN/m.

Ans. A change in temperature results in a change in dip. Hence H changes to H + ΔH

l = 10 m ; d = 10 m ; T = 28°C ; α = 12×10^{-6} per °C ; p = 30 kN/m ; ΔH = ?

Change in horizontal tension,
$$\Delta H = -\frac{3l^2 \alpha T}{16d^2} \times H$$

$$H = \frac{pl^2}{8d} = \frac{30 \times 110^2}{8 \times 10} = 4537.5 \text{ kN}$$

$$\Delta H = -\frac{3 \times 110^2 \times 12 \times 10^{-6} \times 28 \times 4537.5}{16 \times 10^2}$$

$$= -34.589 \text{ kN}$$

∴ Change in horizontal tension, ΔH = – 34.589 kN.

32. What are the main functions of stiffening girders in suspension bridges ?

Ans. Functions of stiffening girders are :

(*i*) Stiffening girders are needed to help the cables keep their shape.

(*ii*) Whatever be the live load on the deck slab, the stiffening girders will convert and transmit the load on the deck slab as a uniformly distributed load to the cables and thereby help the cables retain the parabolic shape during the passage of loads.

(*iii*) The dead load of the girders, being udl, is fully transmitted to the cable and is entirely taken up by the tension in the cable.

(*iv*) The stiffening girders will have to resist a part of the shear forces and bending moments due to live loads.

15

Space Trusses

1. What is a space truss ?

Ans. A frame consisting of members jointed (ball and socket) together at their ends to form a stable three dimensional structure is called a space truss.

2. Give some examples for space trusses.

Ans. Examples for space trusses:

1. Transmission towers
2. Space truss roofs
3. Girders for suspension bridges
4. Steel offshore structures.

3. Differentiate a plane truss and a space truss.

Ans. A truss is an articulated structure composed of straight members arranged and connected in such a way that they transmit primarily axial forces. If all the members lie in one plane it is called a plane truss. A three dimensional truss is called a space truss.

S.No.	*Plane truss*	*Space truss*
1.	All the members lie in one plane.	This is a three dimensional truss.
2.	All joints are assumed to be hinged with freedom to swivel in the X-Y plane.	All joints are assumed to be ball and socketed joints enabling each member to swivel in the X, Y and Z directions.
3.	Inherently unstable in the Z direction.	No stability problem with space trusses.

4. Give the relationship between the number of members and number of joints in a determinate space truss.

Ans. $m = (3j - 6)$ is the relationship between the number of members (m) and number of joints (j) in a determinate space truss.

5. List the assumptions made in the design of space trusses.

Ans. Assumptions:

1. Members are connected together using ball and socket joints

2. Loads and reactions are applied centrally at the joints, and
3. Centroid of members connected at a joint are concurrent.

6. What are the methods available for the analysis of space frames ?

Ans. The methods are:

1. Method of joints
2. Method of tension coefficients
3. Matrix methods
4. Finite element method.

7. What is the minimum number of members required to develop a stable space truss having 4 joints ?

Ans. For a stable space truss,

$$m = (3j - 6) = (3 \times 4 - 6) = 6$$

∴ Minimum 6 members are required to develop a stable space truss.

8. How will you find the internal indeterminacy of space trusses ?

Ans. In a space truss,

If $m + r = 3j \rightarrow$ Determinate truss

$m + r > 3j \rightarrow$ Indeterminate truss

$m + r < 3j \rightarrow$ Unstable truss

where m = number of members

j = number of joints

r = number of support reactions.

9. Check the internal indeterminacy of the space truss shown below:

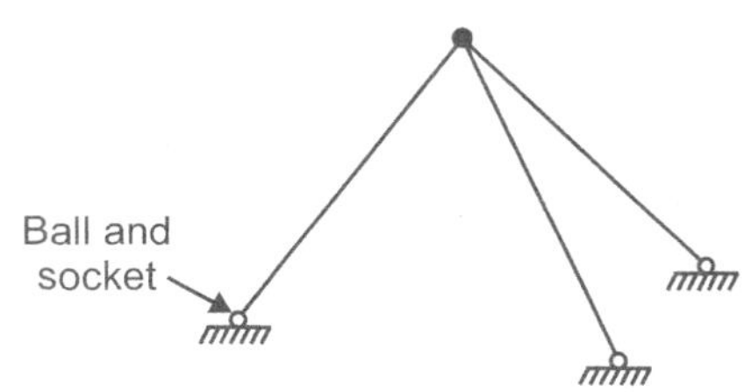

Fig. 1

Ans. $m = 3; j = 4; r = 3 \times 3$

$= 9$ (3 reactions/joint)

$m + r = 3 + 9 = 12$

$3j = 3 \times 4 = 12$

$m + r = 3j$

∴ Determinate truss.

10. What is the least number of support reactions needed for a stable and externally determinate space truss ?

Ans. The least number of support reactions required for a stable and externally determinate space truss is six.

i.e., if $r = 6 \rightarrow$ stable and externally determinate

$r < 6 \rightarrow$ unstable externally

$r > 6 \rightarrow$ externally indeterminate

11. List the types of supports adopted and their reaction components, in space trusses.

Ans.

S. No.	*Type of support*	*Reaction components possible*
1.	Short link	One reaction along the axis of the link
2.	Slotted roller constrained in a cylinder	Two reactions along the two axes perpendicular to the axis of the cylinder
3.	Roller	One reaction perpendicular to the plane on which the roller rests
4.	Ball and socket	Three reactions one each along the coordinate axes

12. A space truss is supported by two short links at A, one short link at B and a ball and socket joint at C. Comment on the external determinacy of the truss.

Ans.

Support	*Type*	*Possible reactions*
A	2 short links	2
B	1 short link	1
C	Ball and socket	3

No. of possible reactions = 6

$$\left.\begin{array}{l}\text{No. of reactions required for stable}\\ \text{and determinate space truss}\end{array}\right\} = 6$$

Hence, externally determinate.

13. A space truss is supported by a slotted roller constrained in a cylinder at A, one short link at B and a ball and socket joint at C. Comment on the external determinacy of the truss.

Ans.

Support	*Type*	*Possible reactions*
A	Slotted roller	2
B	Short link	1
C	Ball and socket	3

No. of possible reactions = 6

$$\left.\begin{array}{l}\text{No. of minimum reactions}\\ \text{required}\end{array}\right\} = 6$$

Hence, externally stable and determinate.

14. A certain member of a loaded indeterminate truss suffers an increase in temperature ? How will this change the force in the member ?

Ans. If it is a compression member the compression will increase. If it is a tension member, the tension will decrease.

15. Define 'tension coefficient' of a truss member.

Ans. The tension coefficient for a member of a truss is defined as the pull or tension in that member divided by its length.

16. In a member of a truss, the force is 10 kN compressive. The length of the member is 2.5 m. Calculate the tension coefficient.

Ans. Tension coefficient, $t = \frac{T}{l}$ (T is negative since the force is compressive)

$$t = \frac{-10}{2.5} = -4.00$$

$$t = -4.00.$$

17. In a particular member of a given truss, the tension coefficient works out to – 4.00 kN/m. The length of the member is 2.5 m. Determine the magnitude and nature of the force in the member.

Ans. Force in the member = tension coefficient × length of the member

$$T_{ij} = t_{ij} \times l_{ij}$$

$$T_{ij} = (-4.00) \times 2.5 = -10.00 \text{ kN}$$

∴ Force in the given member = 10 kN (compressive).

18. For the plane truss shown in Fig. 2, determine the reactions at the supports.

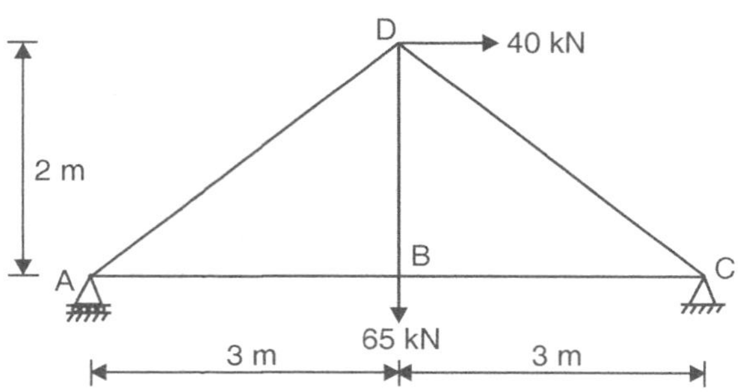

Fig. 2

Ans. $\Sigma H = 0$ gives, $H_C = 40$ kN (←)

Taking moments about C,

$$(V_A \times 6) - (65 \times 3) + (40 \times 2) = 0$$

$$6V_A = (65 \times 3) - (40 \times 2)$$

$$V_A = 19.167 \text{ kN } (\uparrow)$$

$$V_B = 65 - 19.167 = 45.833 \text{ kN } (\uparrow).$$

19. For the plane truss shown in Fig. 2, determine the tension coefficients for the members CB and CD.

Ans.

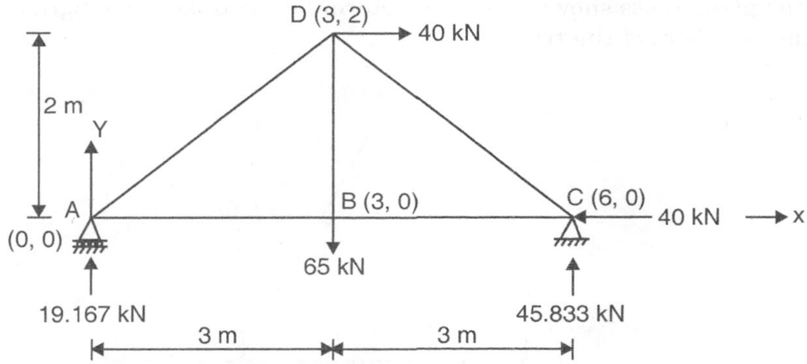

Fig. 3

$$x_{cb} = x_b - x_c = 3 - 6 = -3 \text{ m}$$
$$x_{cd} = x_d - x_c = 3 - 6 = -3 \text{ m}$$
$$y_{cb} = y_b - y_c = 0 - 0 = 0 \text{ m}$$
$$y_{cd} = y_d - y_c = 3 - 0 = 3 \text{ m}$$

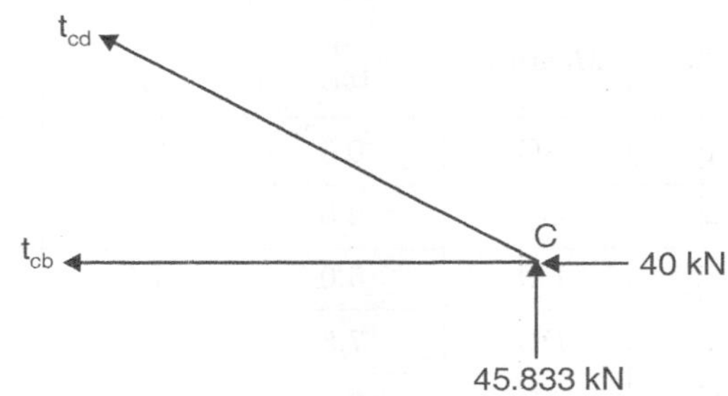

Fig. 4

Joint C :

ΣV = 0 gives,

$$(y_{cd} \times t_{cd}) + 45.833 = 0$$
$$(\uparrow) \qquad (\uparrow)$$
$$3t_{cd} = -45.833$$
$$t_{cd} = -\frac{45.833}{3} = -15.278 \text{ kN/m}$$

ΣH = 0 gives,

$$x_{cd}\, t_{cd} + x_{cb}\, t_{cb} + 40 = 0$$
$$(\leftarrow) \quad (\leftarrow) \quad (\leftarrow)$$
$$(-3)(-15.278) + (-3)\, t_{cb} + 40 = 0$$
$$-3t_{cb} = -3 \times 15.278 - 40$$
$$t_{cb} = 28.611 \text{ kN/m.}$$

20. Using the results available in question 19, determine the magnitude and nature of forces in the members CB and CD.

Ans. We know that,

Force in a member = (tension coefficient × length of the member)

$$T_{ij} = t_{ij} \times l_{ij}$$
$$l_{cb} = \sqrt{x_{cb}^2 + y_{cb}^2} = \sqrt{(-3)^2 + 0^2} = 3 \text{ m}$$
$$l_{cd} = \sqrt{x_{cd}^2 + y_{cd}^2} = \sqrt{(-3)^2 + 3^2} = 4.243 \text{ m}$$
$$T_{cb} = t_{cb} \times l_{cb} = 28.611 \times 3 = 85.833 \text{ kN (tensile)}$$
$$T_{cd} = t_{cd} \times l_{cd} = -15.278 \times 4.243 = -64.825 \text{ kN}$$
$$T_{cd} = -64.825 \text{ kN (compression).}$$

21. For the plane truss shown in Fig. 5, determine the projected lengths along x and y-axes for the members of the truss.

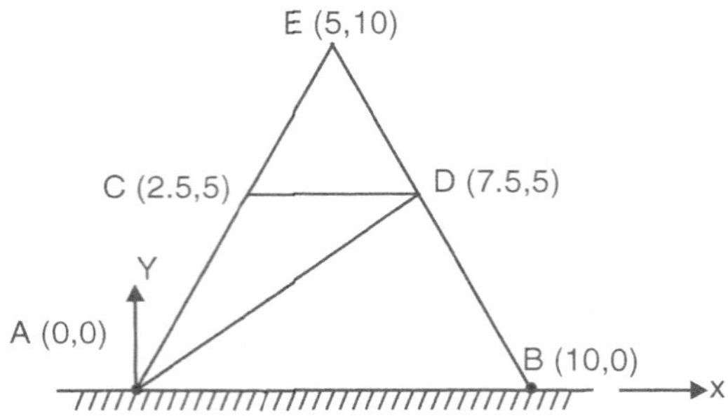

Fig. 5

Ans.

S.No.	Member	x_i (m)	x_j (m)	$x_{ij} = (x_j - x_i)$ (m)	y_i (m)	y_j (m)	$y_{ij} = (y_j - y_i)$ (m)
1.	AC	0.0	2.5	2.5	0	5	5
2.	CE	2.5	5.0	+ 2.5	5	10	5
3.	ED	5.0	7.5	– 2.5	10	5	– 5
4.	DB	7.5	10.0	– 2.5	5	0	– 5
5.	AD	0.0	7.5	7.5	0	5	5
6.	CD	2.5	7.5	5.0	5	5	0

22. From the results of question 21, determine the values of x_{EC}, x_{DC}, y_{CA} and y_{BD}.

Ans. Now the projected lengths in the x and y direction will be calculated in the opposite direction.

$$x_{EC} = -x_{CE} = -(+2.5) = -2.5 \text{ m}$$
$$x_{DC} = -x_{CD} = -(5.0) = -5 \text{ m}$$
$$y_{CA} = -y_{AC} = -(5.0) = -5 \text{ m}$$
$$y_{BD} = -y_{DB} = -(-5.0) = 5 \text{ m.}$$

23. For the plane truss shown in Fig. 5 in question 21, determine the tension coefficients of the members EC and ED if the truss is subjected to a horizontal load of 60 kN at E, towards right.

Ans. $x_{EC} = -2.5$ m ; $y_{EC} = -5$ m

$x_{ED} = 2.5$ m ; $y_{ED} = -5$ m

$$l_{EC} = \sqrt{x_{EC}^2 + y_{EC}^2} = \sqrt{(-2.5)^2 + (-5)^2} = 5.590 \text{ m}$$
$$l_{ED} = \sqrt{x_{ED}^2 + y_{ED}^2} = \sqrt{(2.5)^2 + (-5)^2} = 5.590 \text{ m}$$

Joint E :

$\Sigma V = 0$ gives,

$$(y_{ec} \times t_{ec}) + (y_{ed} \times t_{ed}) = 0$$
$$(\downarrow) \qquad (\downarrow)$$
$$-5t_{ec} - 5t_{ed} = 0$$
$$t_{ec} + t_{ed} = 0$$
$$t_{ec} = -t_{ed}$$

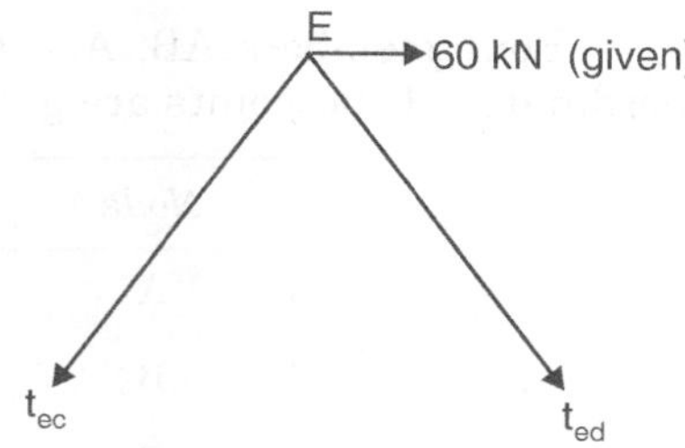

Fig. 6

$\Sigma H = 0$ gives,

$$x_{ec} \cdot t_{ec} + x_{ed}\, t_{ed} + 60 = 0$$
$$(-2.5)(-ted) + 2.5t_{ed} + 60 = 0$$
$$5t_{ed} = -60$$
$$t_{ed} = -12$$

24. At a node A in a space frame (Fig. 7) a force of 1.8 kN acts in the X direction. Locate the member with zero force.

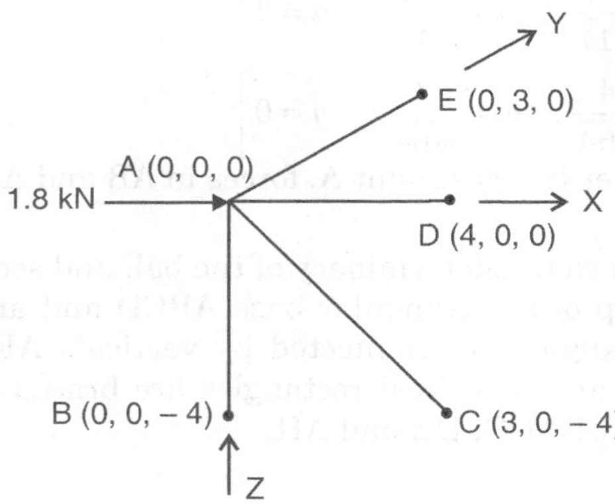

Fig. 7

Ans. AE will have zero force. All others and the external force are in the XZ plane.

25.

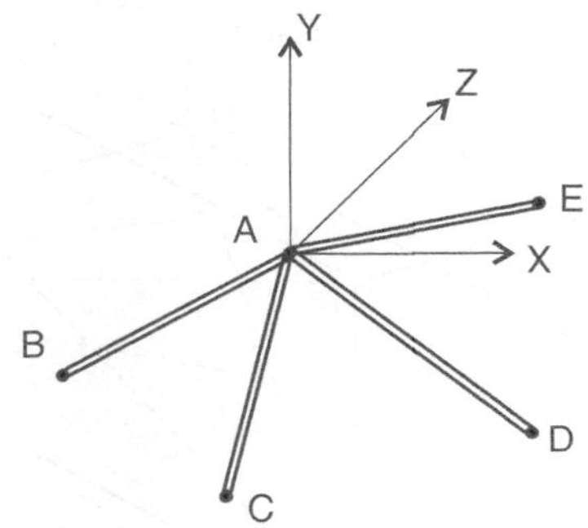

Fig. 8

Four members AB, AC, AD and AE meet at a joint A in a space truss. The spatial coordinates of the points are given below.

Node	*X*	*Y*	*Z*
A	0	0	0
B	– 4	– 1	0
C	1	– 1	– 3
D	4	0	– 2
E	4	4	0

There are no external forces. Members AC and AD have zero forces. Comment on the forces on AB and AE.

Ans. AB and AE are not in the same line since their direction cosines are different.

[For AB, $l = \frac{-4}{\sqrt{17}}$; $m = \frac{-1}{\sqrt{17}}$; $n = 0$

For AE, $l = \frac{4}{\sqrt{64}}$; $m = \frac{4}{\sqrt{64}}$; $n = 0$]

Since there are no other forces at joint A, forces in AB and AE (which are non-collinear) shall be zero.

26. Find the degree of internal indeterminacy of the ball and socket jointed frame in Fig. 9. The frame is made up of a rectangular base ABCD and an identical rectangular top EFGH. These 2 rectangles are connected by verticals AE, BF, CG and DH. The 2 horizontal rectangles and 4 vertical rectangles are braced by diagonals AC, BD, EG, FH, AF, BE, BG, CF, CH, DG, DE and AH.

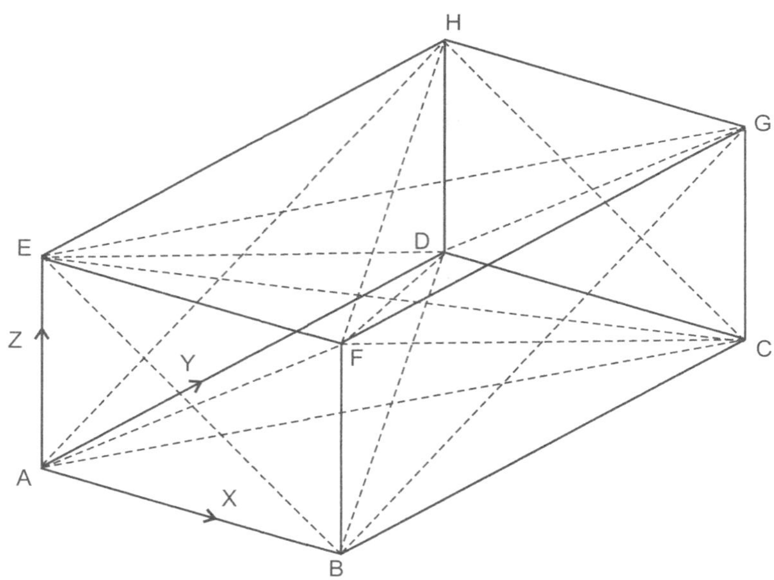

Fig. 9

Ans. Number of members needed for internal stability

$$m = 3j - 6$$

Here, j = 8; actual ma = 4 + 4 + 4 + (6 × 2) = 24

Needed m = (3 × 8) – 6 = 18

Number of redundants = 24 – 18 = 6.

27. How many ground restraints are needed to make the frame in Fig. 9, stable. Give one set of adequate restraints.

Ans. 6 ground restraints are needed to achieve external stability.

These six restraints can be evaluated from the 6 equilibrium equations,

$$\Sigma X = 0, \Sigma Y = 0, \Sigma Z = 0, \Sigma Mx = 0, \Sigma My = 0 \text{ and } \Sigma Mz = 0$$

Sample set of restraints : Prevent displacements along Y, Z at A; Z at B, C and D and X at C.

28. The trestle in Fig. 10 has a ball and socket joint at A.

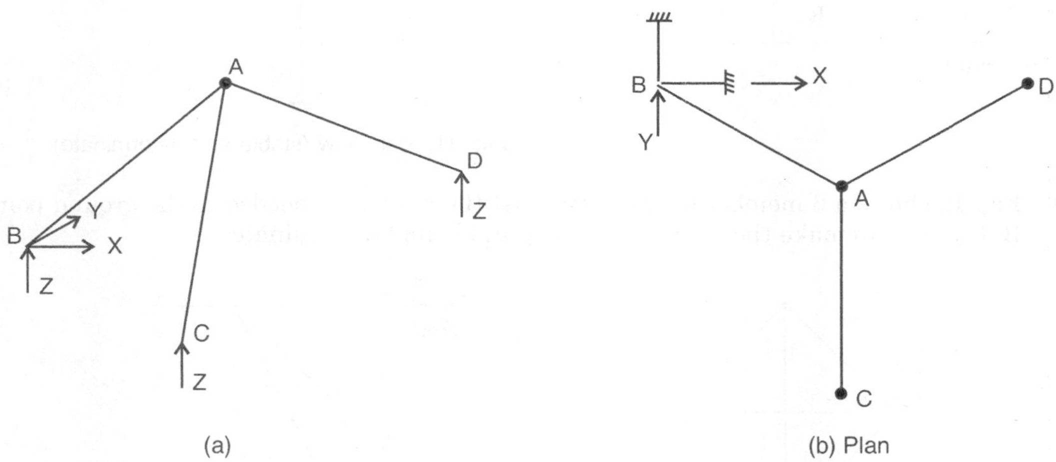

Fig. 10

The following restraints are provided at ground level (BCD).

	X	Y	Z
B	R	R	R
C	Free	Free	R
D	Free	Free	R

(R – restrained)

Obviously the external restraints are inadequate. List the ways in which the frame will collapse.

Ans. 1. C and D can slide outwards.

2. The frame can rotate about Z axis at B.

29. State what restraints are to be added or removed to make the frame in Fig. 10 stable and determinate ?

Ans. The frame itself is deficient. There are 4 joints $j = 4$.

We need $3j - 6 = 6$ members. We have only 3.

For stability there shall be a minimum of 6 external restraints. In addition we must add 3 more members or 3 more restraints. If we add X and Y restraints at C and D, the frame will become stable and determinate.

So we shall restrain the ground points as under (see Fig. 11).

	X	*Y*	*Z*
B	R	R	R
C	R	R	R
D	R	R	R

R–restraint

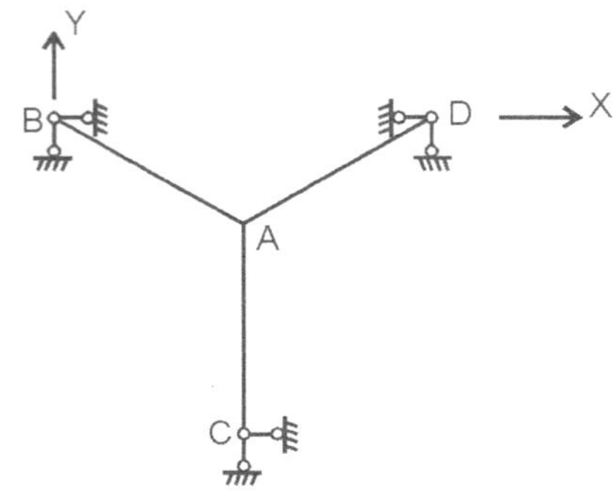

Fig. 11. Plan view (stable and determinate).

30. Fig. 12 shows a 6 member tetrahedron. List the restraints needed to the ground points B, C and D to make the frame externally stable and determinate.

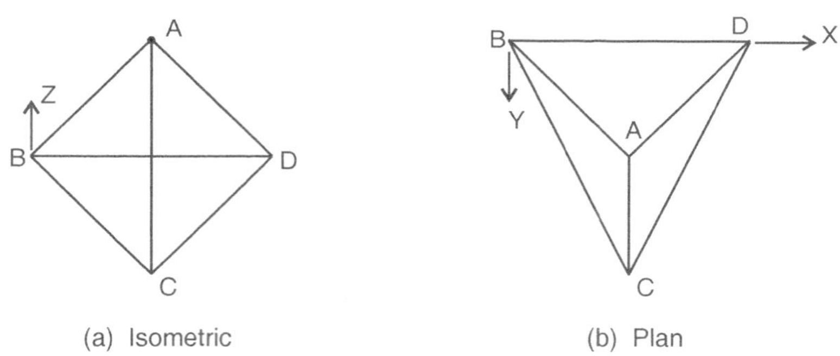

Fig. 12

Ans. For external stability we need a minimum of 6 restraints. They can be provided as under [vide Fig. 12(*c*)].

	X	*Y*	*Z*
B		R	R
C	R		R
D		R	R

R–restraint

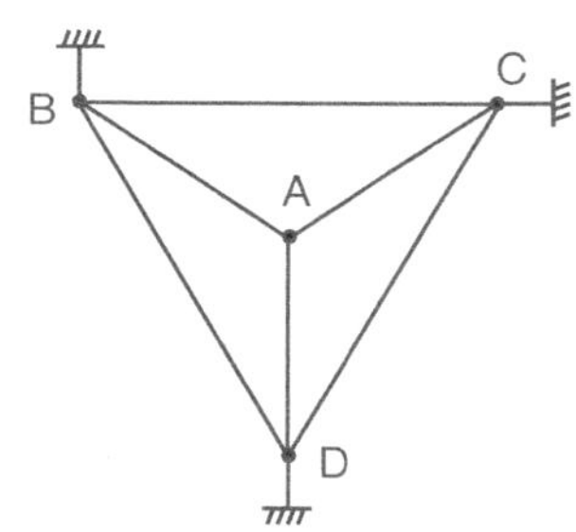

Fig. 12 (c) Plan view.

31. In the space frame shown in Fig. 13, check the internal and external indeterminacies.

Ans. There are 6 external restraints (4 vertical and 2 horizontal). Hence the frame is externally stable and determinate.

There are 20 members (M = 20) and 9 joints (j = 9).

Number of members needed for stability

$$m = 3j - 6 = 27 - 6 = 21$$

The truss is internally deficient. We need one more member for internal stability.

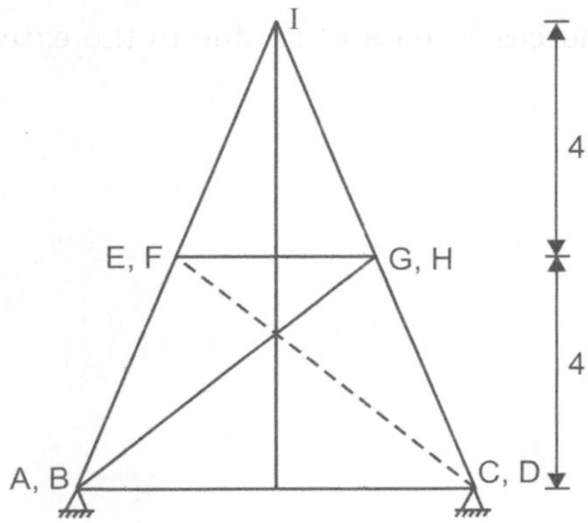

Fig. 13 (a) Elevation.

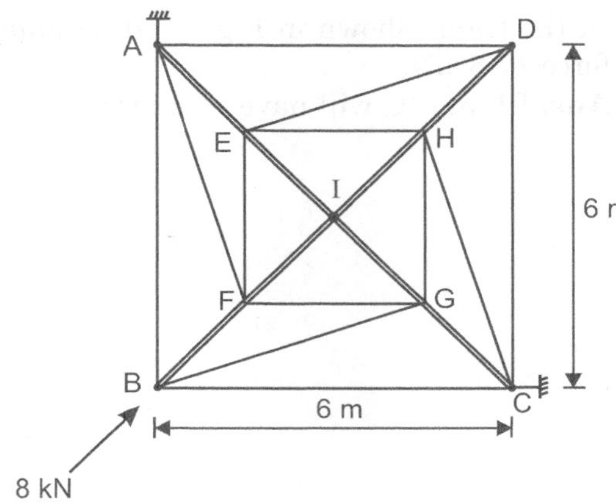

Fig. 13 (b) Plan.

32. For the frame in Fig. 13, suggest a way of making it internally stable.

Ans. Add one more member to make it stable. This can be added along AC or BD or EG or FH.

33. In the frame shown in Fig. 13, find the forces on IF and IH due to an external load of 8 kN as shown. The force is horizontal and acts on the apex I in the diagonal direction.

Ans.

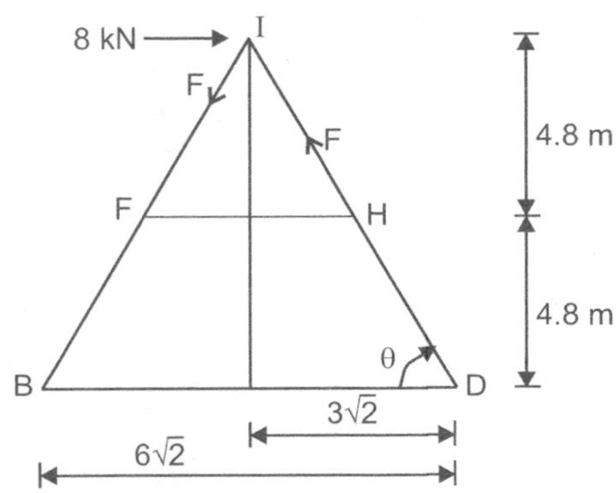

Fig. 14

Since the force is in the diagonal direction the problem reduces to a 2D case.

$$\tan \theta = \frac{9.6}{3\sqrt{2}} = 2.263$$

$$\theta = 66.16°$$

$$2F \cos \theta = 8$$

$$F = \frac{4}{\cos \theta} = 9.9 \text{ kN}$$

Force in IH = 9.9 kN (compression)

Force in IF = 9.9 kN (tension).

34. In the frame shown in Fig. 13, determine the forces in IG and IE due to the external force of 8 kN.

Ans. EI and IG will have zero forces.

16

Beams Curved in Plan

1. Give two examples for beams curved in plan.
Ans. Examples of beams curved in plan are,
(*i*) Beams in a bridge negotiating a curve
(*ii*) Ring beams supporting water tanks, silos, etc.
(*iii*) Beam supporting corner lintels, curved balconies, etc.
(*iv*) Beams in ramps in traffic interchanges.

2. What are the forces developed in a beam curved in plan ?
Ans. Forces developed in a beam curved in plan :
(*i*) Bending moments
(*ii*) Shears and
(*iii*) Torsional moments.

3. A curved beam (in plan) carrying a vertical concentrated load at the free end is shown in Fig. 1. Determine the bending moment and twisting moment at the section X.

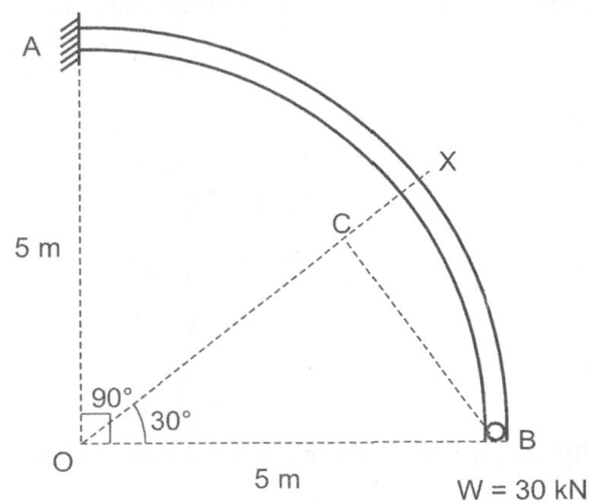

Fig. 1

Ans. $CX = R(1 - \cos\theta) = 5(1 - \cos 30°) = 0.67$ m

$CB = R \sin\theta = 5 \sin 30° = 2.50$ m

Bending moment at the section X,

$$M_\theta = -W(CB) = -30 \times 2.50 = -75 \text{ kNm}$$

Twisting moment at the section X,

$$T_\theta = -W(CX) = -30 \times 0.67 = -20 \text{ kNm}$$

4. For the beam in question 3, sketch the shear force diagram.

Ans.

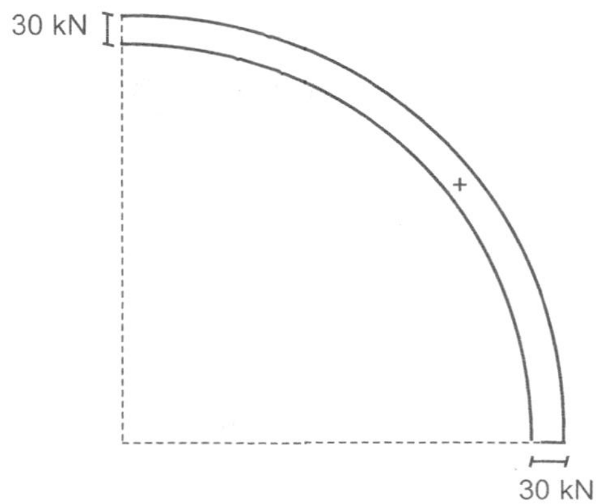

Fig. 2. Shear force diagram.

Shear force at the section X, $F_\theta = W = 30$ kN

F_θ is independent of θ and is uniform throughout. Fig. 2 shows the shear force diagram.

5. List out the forces that must be introduced at the cut face of a beam curved in plan in order to make a proper free body diagram.

Ans. A *free body diagram* must indicate on the cut section all the internal forces that would have been acting at the section before it was cut.

Accordingly Fig. 2(*a*) shows the forces to be introduced at the cut face.

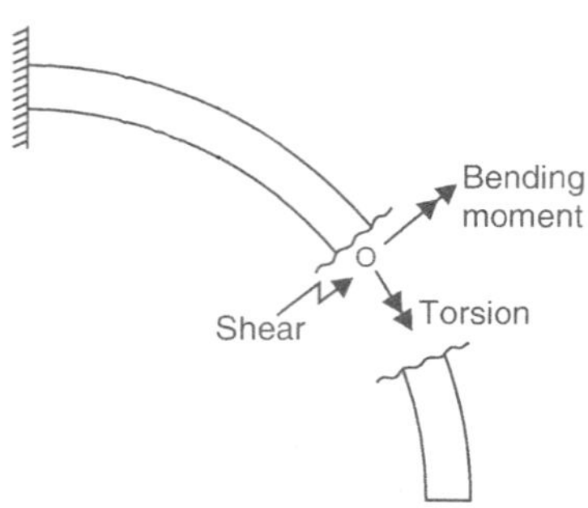

Fig. 2. (a) Free body of a curved beam (plan).

6. A beam curved in plan is loaded at the free end as shown in Fig. 3. Sketch the bending moment diagram.

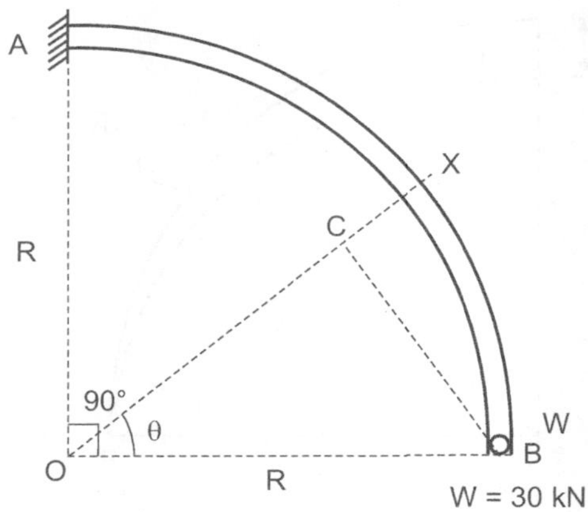

Fig. 3

Ans. Bending moment at section $X = M_\theta = -W(CB) = -WR\sin\theta$

When $\theta = 0$, $M_B = -W \times R \times 0 = 0$

When $\theta = \pi/2$, $M_B = -W \times R \times 1 = -WR$

Fig. 3 (*a*) shows the bending moment diagram.

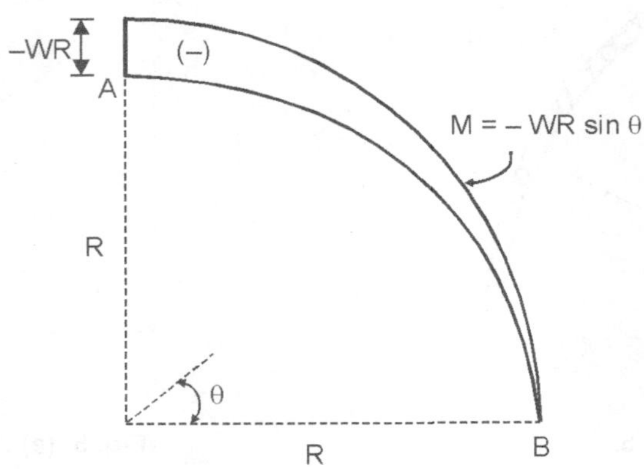

Fig. 3. (a) Bending moment diagram.

7. For the beam curved in plan, in question 6, sketch the twisting moment diagram.

Ans. From Fig. 3, the twisting moment at the section,

$$T_\theta = -W(CX)$$
$$= -W \times R(1 - \cos\theta)$$

When $\theta = 0$, $T_B = -W \times R(1 - \cos\theta) = 0$

When $\theta = \pi/2$, $T_A = -W \times R(1 - \cos\pi/2) = -WR$

Fig. 4 shows the twisting moment diagram.

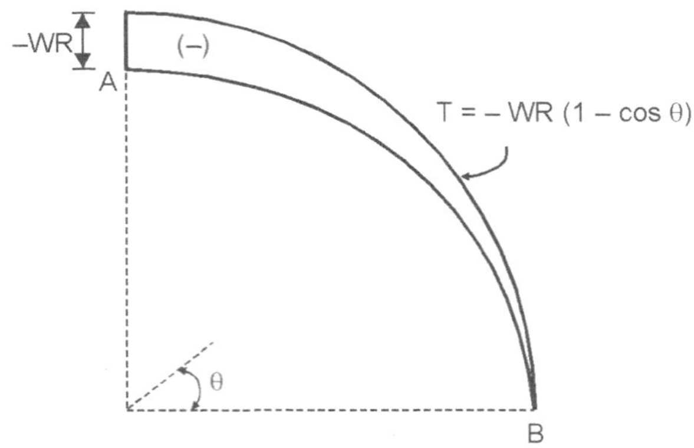

Fig. 4. Twisting moment diagram.

8. A beam curved in plan is subjected to a udl over the entire span as shown in Fig. 5. Sketch the shear force diagram.

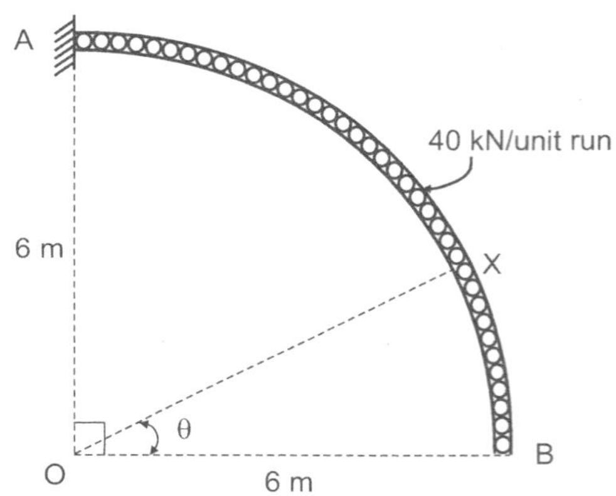

Fig. 5

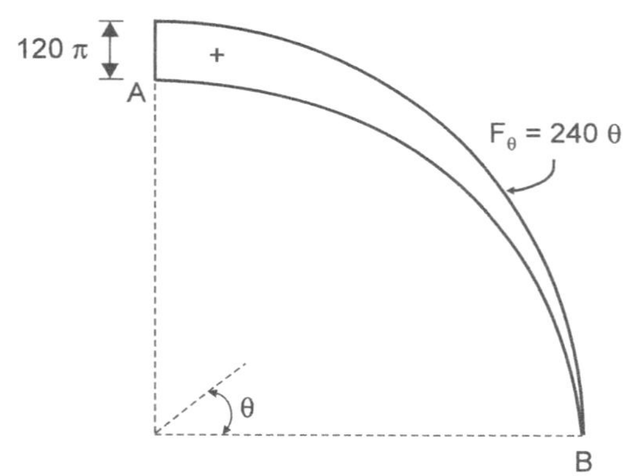

Fig. 5. (a) Shear force diagram.

Ans. Shear force at the section X,

$$F_\theta = w \times R \times \theta = 40 \times 6 \times \theta$$

$$F_\theta = 240\,\theta, \quad (\theta \text{ is in radians})$$

When $\theta = 0$, $F_B = 0$

When $\theta = \pi/2$, $F_A = 240 \times \pi/2$

The variation of shear force is shown in Fig. 5 (*a*).

9. A curved beam semicircular in plan on three equally spaced supports is loaded as shown in Fig. 6. Calculate the reactions at the supports.

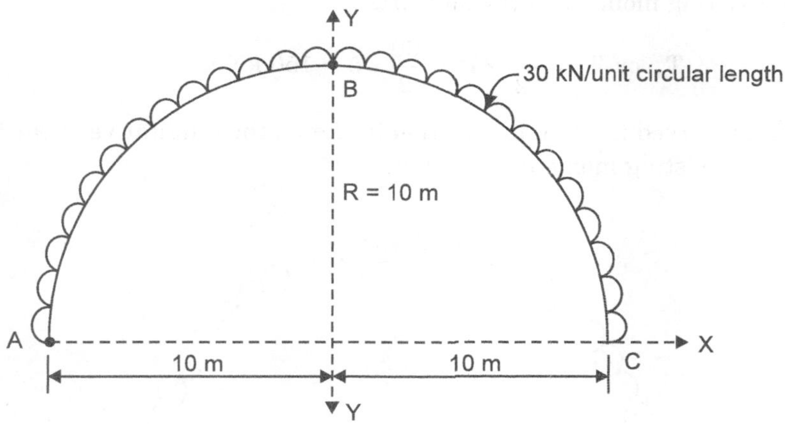

Fig. 6

Ans. The structure, loading and support conditions are symmetric with respect to the YY-axis.

Hence, $V_A = V_C$.

Taking moments of all forces about a tangent at point B,

$$2\,(V_A \times 10) - 30 \times \pi \times 10\left(R - \frac{2R}{\pi}\right) = 0$$

$$20V_A = 30 \times \pi \times 10\left(10 - \frac{2 \times 10}{\pi}\right)$$

$$20V_A = 3424.78$$

$$V_A = 171.24 \text{ kN} = V_C$$

$$V_B = \text{Total load} - 2V_A$$

$$= 30 \times \pi \times 10 - 2\,V_A = 300\pi \times 2 \times 171.24 = 600 \text{ kN}.$$

10. For the bow girder shown in Fig. 7, find the twisting moments at the supports A and B due to the 60 kN load.

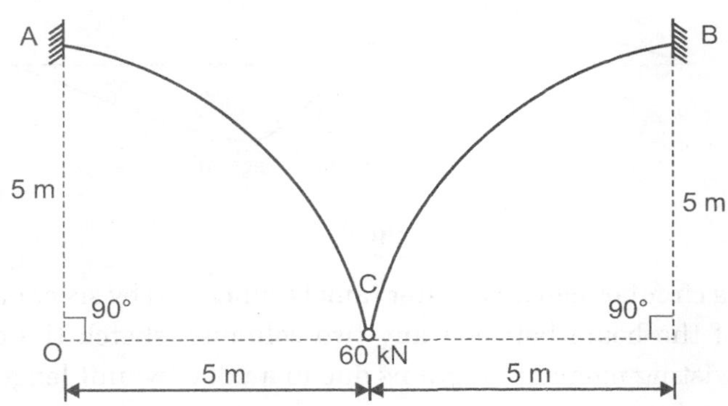

Fig. 7

Ans. The twisting moment at the supports

$$T_A \text{ or } T_B = \frac{W}{2} \times R = \frac{60}{2} \times 5 = 150 \text{ kNm.}$$

11. For the beam curved in plan, in question 9, sketch the qualitative shear force, bending moment and twisting moment diagrams.

Ans.

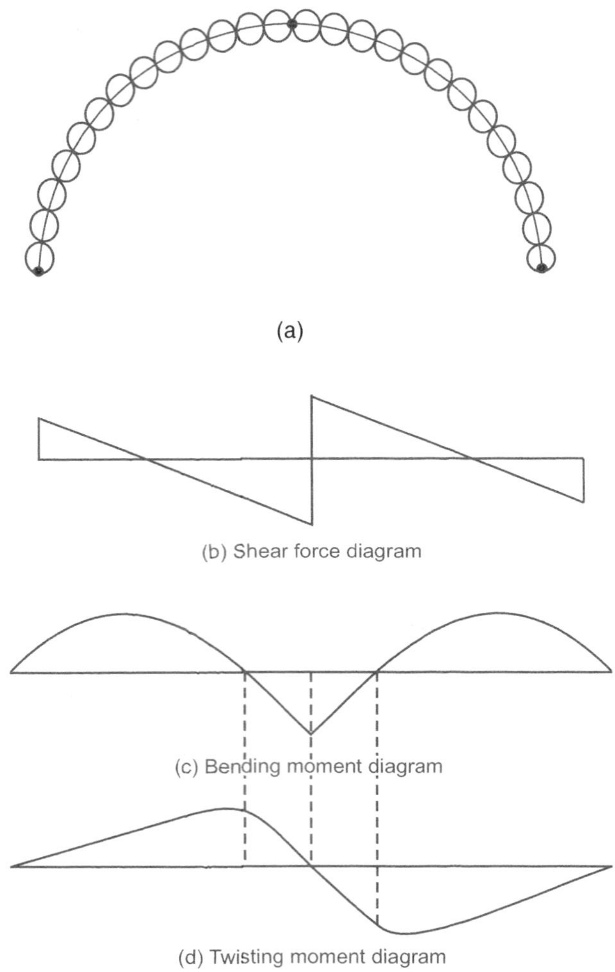

(a)

(b) Shear force diagram

(c) Bending moment diagram

(d) Twisting moment diagram

Fig. 8

12. A ring beam of a circular, elevated water tank is supported by six equally spaced columns. For the part of the beam between any two columns, sketch the qualitative bending moment and twisting moment diagrams due to a udl of w/unit length on the ring beam.

Ans.

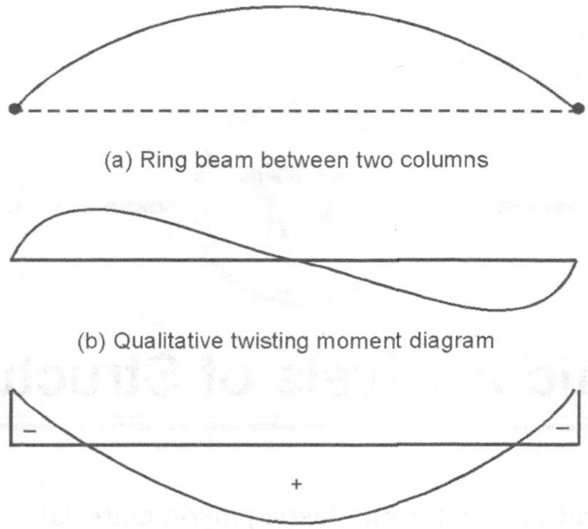

(a) Ring beam between two columns

(b) Qualitative twisting moment diagram

(c) Qualitative bending moment diagram

Fig. 9

13. In a circular beam on several equally spaced supports, what are the quantities which will be zero at the supports and why ?

Ans.

1. Slope on either side of any support will be zero. This is because a diametral line through any column is a line of symmetry.

2. Torsional moment will be zero at every support. This is because, any torque at a support would make it unsymmetrical.

14. A corner lintel is built into the walls so that A and C (vide Fig. 10) can be treated as fixed. Find the bending moment and torsion at A due to a point load of 4.8 kN at B.

Ans. Moment about line AC,

$M_{AC} = 4.8 \times 0.8 \cos 45° = 2.7153$

Because of symmetry about BD,

M_{AC} will be shared equally at A and C

$$\text{Moment at A} = \frac{2.7153}{2} = 1.3576$$

Component along AB (torque)

$= 1.3576 \cos 45° = 0.96$ kNm

Component normal to AB (BM) = 0.96 kNm

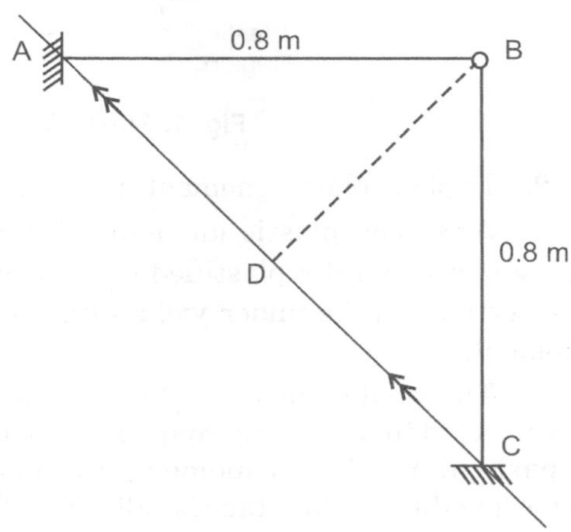

Fig. 10

17

Plastic Analysis of Structures

1. Draw the stress-strain curve for an elastic-plastic material.

Ans.

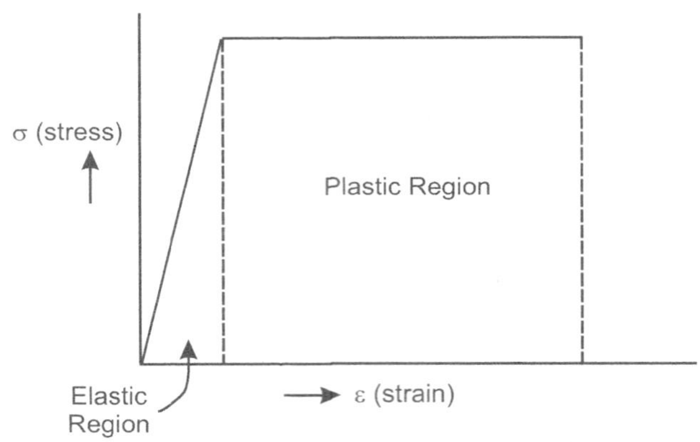

Fig. 1. Stress-Strain curve for elastic-plastic material.

2. Explain plastic moment and yield moment.

Ans. The plastic moment (M_p) of a section is defined as the maximum moment of resistance of a fully plastified or yielded cross-section. Here, all fibres from top to bottom of the section will be under yield stress and the section would not be able to take any further moment.

The yield moment (M_y) is the moment at which the first yield occurs. When the section is subjected to increasing couples, the stress in the cross-section will gradually increase and at a particular value of moment, the most distant fibre/s will start yielding. The moment corresponding to this stage is called the yield moment. Figs. 1(*a*) and (*b*) show the state of stress in a cross-section at plastic moment and yield moment for a section symmetrical about *x*-axis.

3. Define shape factor.

Ans. Shape factor (S) of a beam is defined as the ratio of the plastic moment of a section to the yield moment of the section.

$$S = \frac{M_p}{M_y} = \frac{\sigma_y Z_p}{\sigma_y Z} = \frac{Z_p}{Z} = \frac{\text{Plastic modulus of section}}{\text{Elastic section modulus}}$$

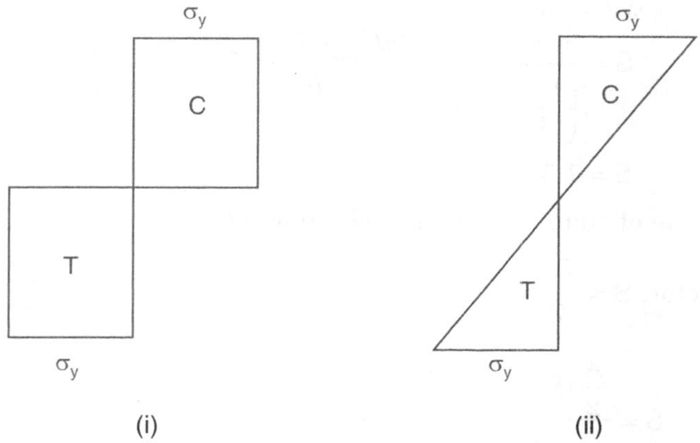

Fig. 1 (a) States of stress at (i) plastic moment and (ii) yield moment.

(Shape factor is also the ratio of plastic modulus of the section to the elastic modulus of the section.)

4. Distinguish between plastic modulus and section modulus.

Ans. The plastic modulus of a section Z_p is the first moment of the areas above and below the equal area axis. It is the resisting modulus of the completely plastified section.

The section modulus of a section is defined as the moment of inertia about the neutral axis divided by the distance of the most distant point of the section from the neutral axis.

5. State the lower bound theorem for collapse load.

Ans. The lower bound theorem states that "A load computed on the basis of an assumed equilibrium bending moment diagram in which the moments are nowhere greater than M_p is less than or at best equal to the true ultimate load".

6. State the upper bound theorem for collapse load.

Ans. The upper bound theorem states that "A load computed on the basis of an assumed mechanism is always greater than or at best equal to the true ultimate load".

7. What is a plastic hinge ? Explain.

Ans. A plastic hinge is defined as a yielded zone due to bending in a structural member at which large rotations can take place at a section at a constant plastic moment, M_p.

Fully plastic moment is considered to have developed at any section of a structure subjected to a system of loads when the section is completely yielded or plastified. The fibres on one side of the equal area axis of the section are in compression and tend to contract. The fibres on the other side of the axis are in tension and tend to extend. The section acts like a hinge. This hinge is known as a plastic hinge.

8. What is the shape factor for a rectangular section ?

Ans. Shape factor, $S = \dfrac{\text{Plastic section of modulus}}{\text{Elastic section modulus}}$

$$S = \frac{\frac{A}{2}(\bar{y}_1 + \bar{y}_2)}{\frac{I}{y}} = \frac{\frac{bd}{2}\left(\frac{d}{4} + \frac{d}{4}\right)}{\left(\frac{bd^3}{12}\right) \Big/ \left(\frac{d}{2}\right)}$$

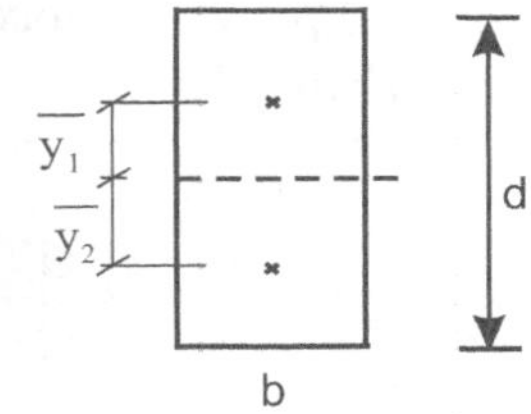

Fig. 2

$$S = \frac{\left(\frac{bd}{2} \times \frac{d}{2}\right)}{\left(\frac{bd^3}{12} \times \frac{2}{d}\right)} = \frac{bd^2}{4} \times \frac{6}{bd^2} = 1.5$$

S = 1.5.

9. What is the value of shape factor for a circular section ?

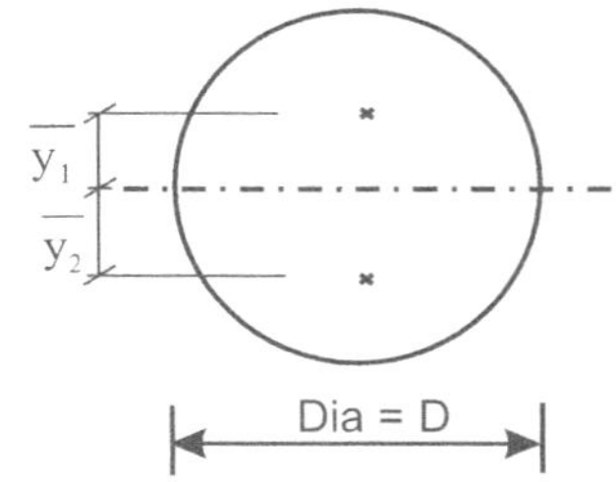

Fig. 3

Ans. Shape factor, $S = \dfrac{Z_p}{Z}$

$$S = \frac{\frac{A}{2}(\bar{y}_1 + \bar{y}_2)}{\frac{I}{y}}$$

$$S = \frac{\frac{\pi}{4} D^2 \times \frac{1}{2}\left(\frac{2D}{3\pi} + \frac{2D}{3\pi}\right)}{\left(\frac{\pi}{64} D^4\right) \Big/ \left(\frac{D}{2}\right)}$$

$$S = \frac{\pi}{4} D^2 \times \frac{1}{2} \times \frac{4D}{3\pi} \times \frac{64}{\pi D^4} \times \frac{D}{2} = 1.697$$

S = 1.697

10. Define collapse load of a flexural frame.

Ans. A flexural frame when subjected to steadily increasing loads develops plastic hinges at one or more locations. When enough hinges are formed to reduce the frame to a mechanism, collapse will occur. The load causing this collapse is obviously called the collapse load.

11. What is the relation between the static indeterminacy of a frame and the number of plastic hinges needed to cause collapse?

Ans. A frame with n degrees of indeterminacy will require $n + 1$ plastic hinges to get formed to become a (collapse) mechanism.

12. List the possible locations of plastic hinges in a structure.

Ans. The plastic hinges are likely to be formed:

(*i*) at the points under concentrated loads

(*ii*) points of supports

(*iii*) points of connections or joints

(*iv*) fixed end supports

(*v*) points of maximum bending moment.

Example:

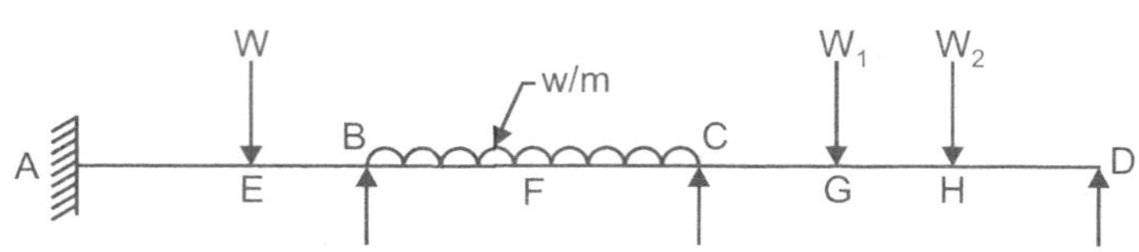

Fig. 4

Plastic hinges possible at: A, B, C, D, E, F, G, H → (7)

Structural hinge at D.

13. How many plastic hinges are needed for the formation of a mechanism in a propped cantilever ?

Ans. A propped cantilever (with vertical loading only) is statically indeterminate to 1st degree. On the formation of one plastic hinge, the structure will act as a simply supported beam. On the formation of one more hinge the structure will become a mechanism.

Hence, two plastic hinges are needed for the formation of a mechanism in a propped cantilever.

14. What is a mechanism ?

Ans. When an n degree statically indeterminate structure is loaded, on the formation of n plastic hinges, it becomes determinate. It will continue to behave as a structure. On further loading when one more hinge is formed, the structure collapses *i.e.*, on the formation of $(n + 1)$ plastic hinges, the structure will not be able to take any further load and that is called "mechanism" condition.

15. What is meant by fully plastic condition in a structure ?

Ans. Fully plastic condition in a structure means that sufficient plastic hinges are formed to transform the structure into a mechanism.

16. Determine the collapse load for a simply supported beam with a central point load ?

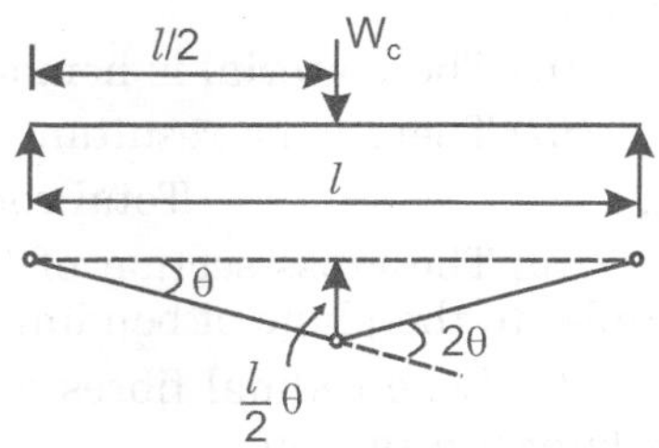

Fig. 5

Ans. External work done = Internal work done

$$W.\frac{l}{2}\theta = 0 + M_p\, 2\theta + 0$$

$$W_c = 2\,M_p\,\theta \times \frac{2}{l\theta} = \frac{4M_p}{l}$$

Collapse load, $W_c = \dfrac{4M_p}{l}$, where M_p is the plastic moment of the beam section.

17. Define 'load factor' of a structure.

Ans. Load factor of a structure is defined as the ratio of the collapse load to the working load, $Q = (P_c/P)$

Load factor is also defined as the product of shape factor and the factor of safety.

$$Q = S\,(F.S).$$

18. Distinguish between 'safety factor' and 'load factor'.

Ans. 'Safety factor' is the term employed in elastic designs. It is the ratio of the yield stress to working stress.

$$SF = \frac{\sigma_y}{\sigma_w}$$

'Load factor' is the term employed in ultimate load design or limit state design.

It is the ratio of the collapse load to the working load.

$$Q = \frac{W_c}{W}$$

19. Sketch the stress distribution in a fully plastic rectangular section.

Ans.

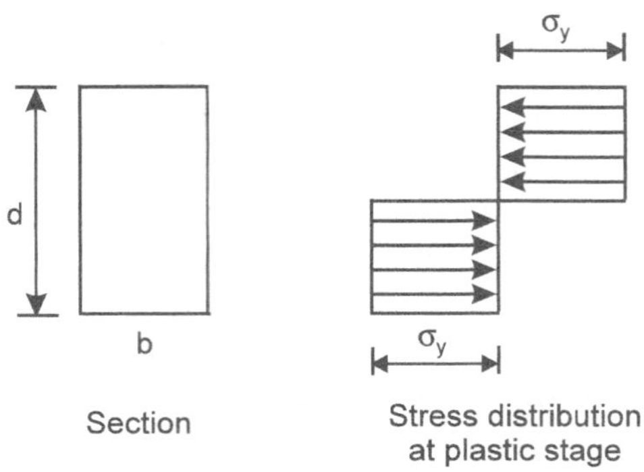

Fig. 6

20. What are the assumptions made in plastic theory ?

Ans. Assumptions :

(*i*) Plane transverse sections remain plane and normal to the longitudinal axis after bending.

(*ii*) The material is homogenous and isotropic in both the elastic and plastic state.

(*iii*) There is no resultant axial force on the cross-section.

i.e., Total compression = Total tension.

(*iv*) The cross-section of the beam is symmetrical about an axis through its centroid parallel to the plane of bending.

(*v*) Longitudinal fibres are free to expand and contract without affecting the fibres in the lateral direction.

21. How will you determine the number of independent mechanisms for a statically indeterminate structure ?

Ans. The number of independent mechanisms (n) is obtained using the relation,

$$n = N - r,$$

where N is the number of possible plastic hinges, (plastic hinges are likely to be formed at the points under concentrated loads, points of supports, points of connections or joints or points of maximum bending moment) and 'r' is the degree of redundancy of the structure.

22. What are the conditions to be satisfied for the plastic method of analysis ?

Ans. Three fundamental conditions are to be satisfied in the plastic analysis :

(*i*) **Mechanism condition.** The ultimate or collapse load is reached when a mechanism is formed. The number of plastic hinges developed should be just sufficient to form a mechanism.

(*ii*) **Equilibrium condition**

$$\Sigma V = 0 \; ; \quad \Sigma H = 0 \quad \text{and} \quad \Sigma M = 0$$

(*iii*) **Plastic moment condition.** The bending moment at any section of the structure should not be more than the fully plastic moment of the section.

23. Explain redistribution of moments (or moment redistribution).

Ans. When an indeterminate structure is subjected to a constantly increasing load, one or more sections get plastified. The plastified sections can not take any additional bending moment. If the increase in load continues, these sections will act as hinges. The behaviour of the structure at this stage would be different. The plastic hinges will now rotate like hinges and allow the additional bending moment to be taken by some other part of the frame. This is called redistribution of moments.

For example, in a 2 span continuous beam shown in Fig. 7, '*a*' is the continuous beam ; it carries a load causing M_p at B. '*b*' is the bending moment diagram for '*a*'. '*c*' is the beam with the hinge at B. It can not take any more bending moment at B. When '*c*' takes additional loads, the bending moment diagram will be as in '*d*'. Here the bending moment due to additional loads has been taken by the spans AB and BC as if they were SS beams.

In other words, the bending moment that would have been taken by section B if it were elastic, has been given away to the spans AB and BC.

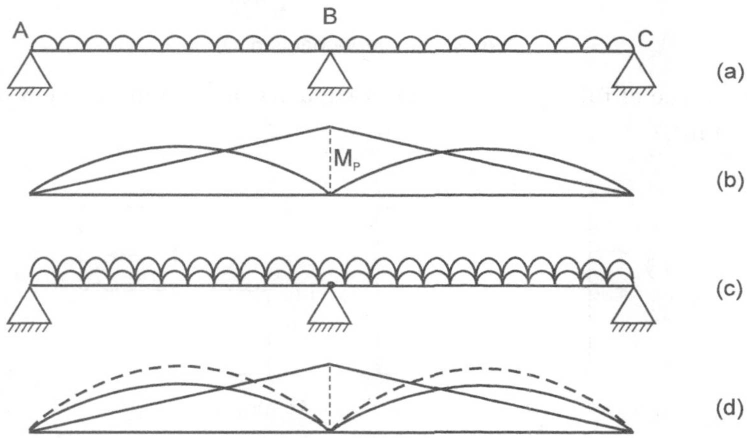

Fig. 7

This is the concept of redistribution of moments.

24. What are the mechanisms to be considered in the plastic analysis of structures ?

Ans. Types of mechanisms : Independent and combined mechanisms are to be considered.

Mechanism

- Independent mechanisms
  - (1) Beam mechanism
  - (2) Panel or sway mechanism
  - (3) Gable mechanism
  - (4) Joint mechanism
- Combined or composite mechanisms

25. What is the length of a plastic hinge of a simple beam of span l subjected to a udl on its entire length ?

Ans. When a simply supported beam of span l is subjected to udl on its entire length, the length of the plastic hinge is given by

$$l_p = \frac{l}{\sqrt{3}} \quad \text{(if a rectangular cross-section is adopted).}$$

This is the length of a beam which has reached yield stress at top and bottom fibres, when the mid section is fully plastic.

26. Sketch the panel mechanism in the case of a portal frame with fixed ends at the base of both columns.

Ans.

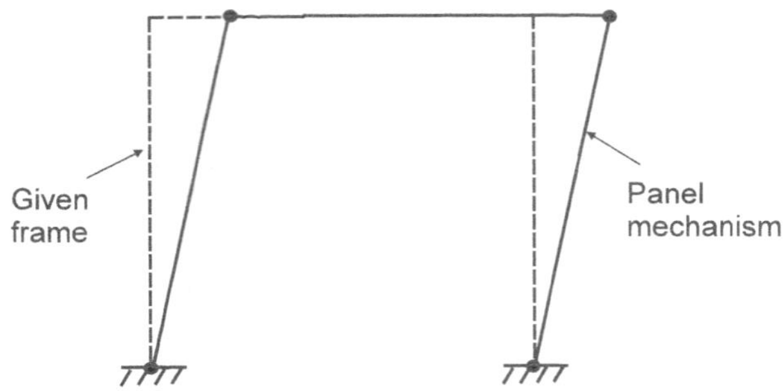

Fig. 8. Panel mechanism.

27. What is the value of fully plastic moment capacity of a square section of side 60 mm. $\sigma_y = 240$ N/mm^2.

Ans.

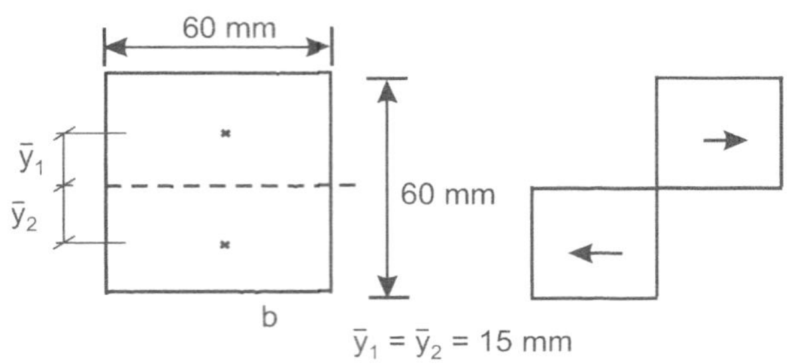

Fig. 9

Plastic moment capacity, $M_p = \sigma_y Z_p$

$$Z_p = \frac{A}{2}(\bar{y}_1 + \bar{y}_2) = \frac{60 \times 60}{2}(15 + 15) = 54000 \text{ mm}^3$$

$$M_p = 240 \times 54000 = 12.96 \times 10^6 \text{ Nmm} = 12.96 \text{ kNm}$$

Plastic moment capacity, $M_p = 12.96$ kNm

28. Find the plastic moment needed in a fixed beam to support an ultimate central concentrated of P.

Ans. External work done = Internal work done

$$P \cdot \frac{l}{2}\theta = M_p\theta + M_p 2\theta + M_p\theta$$

$$P \cdot \frac{l}{2}\theta = 4M_p\theta$$

$$\frac{Pl}{2} = 4M_p$$

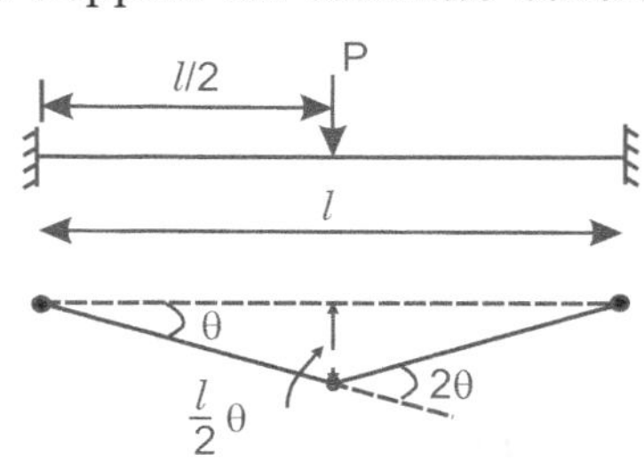

Fig. 10. Mechanism.

Plastic moment, $M_p = \dfrac{Pl}{8}$

29. Find the shape factor of a box section of outer dimensions 80 × 60 mm and inner dimensions 60 × 40 mm

Ans.

$$Z = \frac{\frac{1}{12}(BD^3 - bd^3)}{40}$$

$$= \frac{60 \times 80^3 - 40 \times 60^3}{12 \times 40}$$

$$= 46000 \text{ mm}^3$$

$$Z_p = \frac{A}{Z}(\bar{y}_1 + \bar{y}_2)$$

$$= \frac{2 \times 60 \times 10}{2}(35 + 35) + \frac{2 \times 30 \times 10 \times 2}{2}(15 + 15)$$

$$= 60 \times 10 \times 70 + 30 \times 10 \times 2 \times 30$$

$$= 60{,}000 \text{ mm}^3$$

$$SF = \frac{Z_p}{Z} = 1.3043.$$

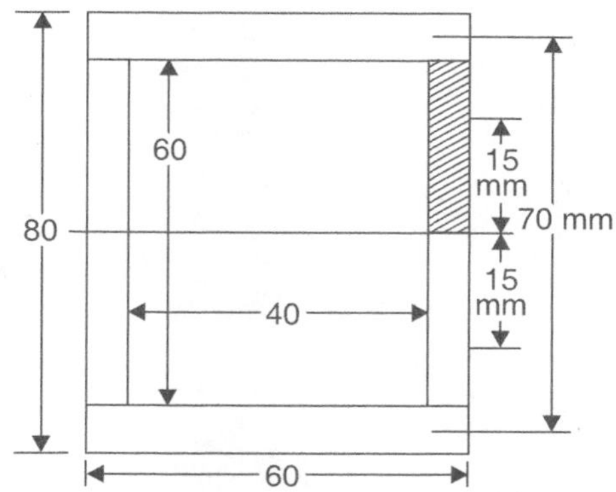

Fig. 11

30. What is the length of plastic hinge of a simple beam of span l subjected to a central concentrated load W ?

Ans. When a simply supported beam of span l is subjected to a central concentrated load W, the length of the plastic hinge,

$l_p = \dfrac{l}{3}$ if a rectangular section is used and

$l_p = \dfrac{l}{8}$ if an I section is used for the same.

31. A fixed beam has a stepped section (Fig. 12). Find the energy dissipated at the hinges if the hinges C and D move down by 0.03 m due to loads W, W at C and D. The plastic moments are as indicated in figure.

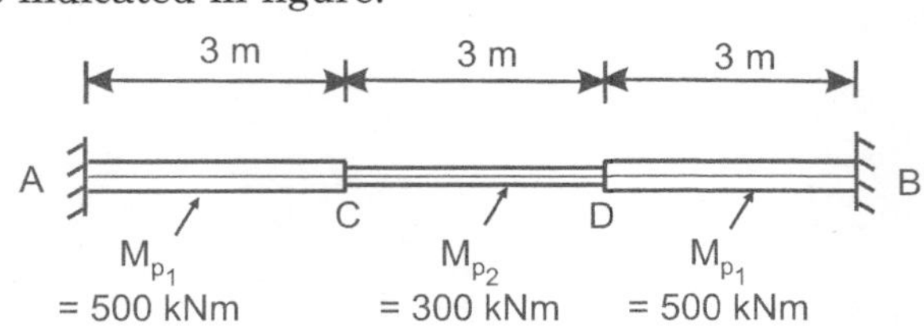

Fig. 12

Ans.

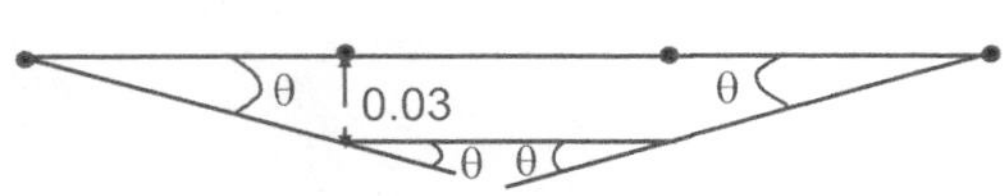

Fig. 12 (*a*)

Referring to Fig. 12 (*a*), $\theta = 0.03/3.0 = 0.01$ rad.

Energy dissipated at the 4 hinges, $E = 2\ M_{p_1}\ \theta + 2\ M_{p_2}\ \theta$

$$= 2 \times 500 \times 0.01 + 2 \times 300 \times 0.01 = 16 \text{ kNm}$$

32. For the frame shown in Fig. 13, indicate the possible failure mechanisms.

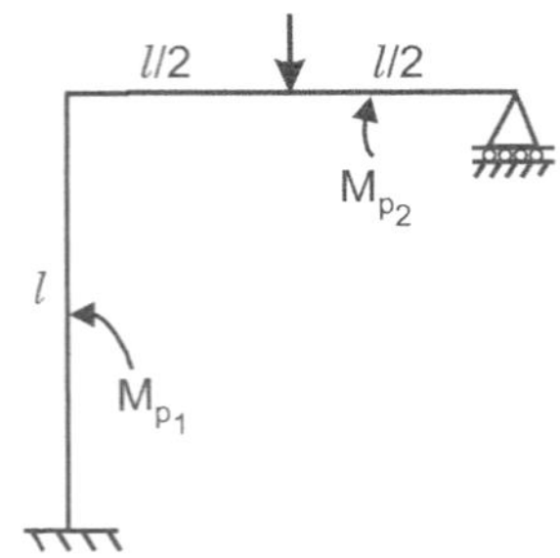

Fig. 13

Ans.

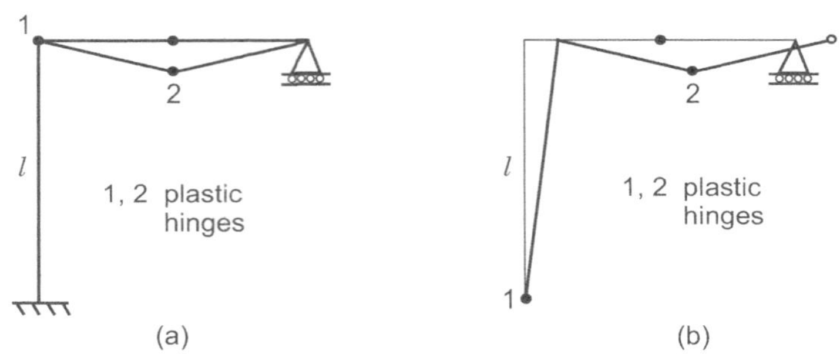

Fig. 14

18

Redundant Frames

1. Find the degree of internal indeterminacy of the frame shown in Fig. 1.

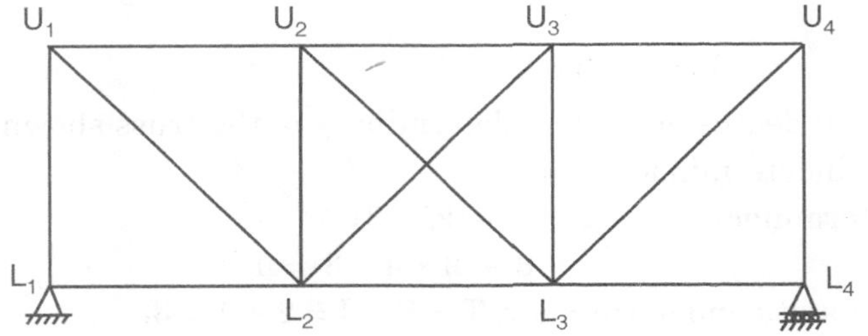

Fig. 1

Ans. Degree of internal indeterminacy,

$$I = m - (2j - 3)$$
$$m = \text{No. of members} = 14$$
$$j = \text{No. of joints} = 8$$
$$\therefore \quad I = 14 - (2 \times 8 - 3) = 14 - 13 = 1$$
$$I = 1.$$

2. How will you make the frame shown in Fig. 1 internally determinate ? Sketch the same.

Ans. Since the above frame is internally indeterminate to one degree, by removing a member like U_2L_3 or L_2U_3, the structure can be made determinate. The resulting structure is sketched below (Fig. 2).

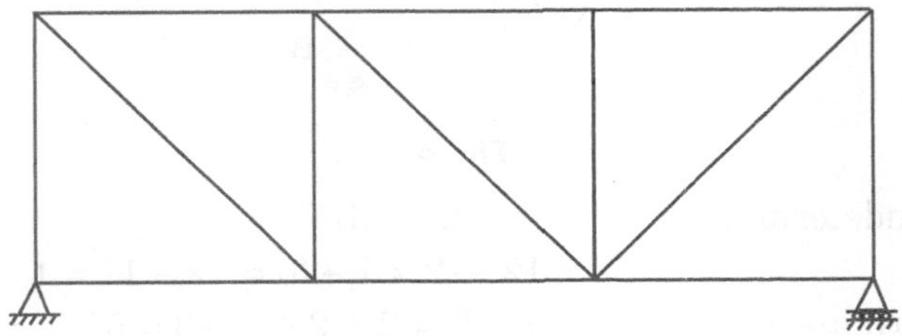

Fig. 2. Statically determinate.

3. Determine the degree of external indeterminacy of the frame shown in Fig. 3.

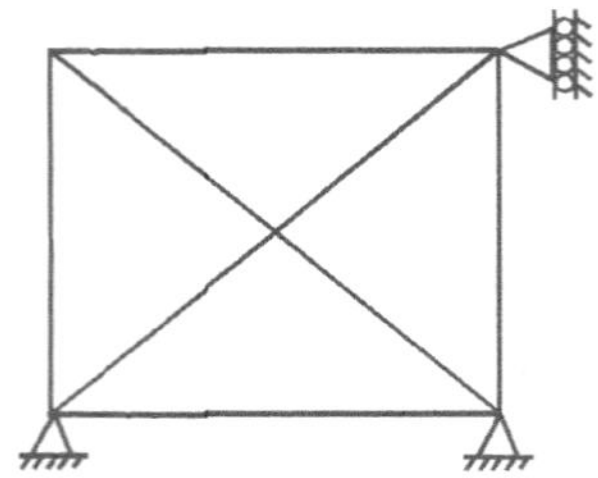

Fig. 3

Ans. External indeterminacy, E = r – R

r = No. of reaction components possible/available
= 5
R = No. of condition equations available
= 3

∴ E = 5 – 3 = 2.

4. What is the total degree of static indeterminacy of the truss shown in Fig. 3?

Ans. External indeterminacy, E = 2

Internal indeterminacy, $I = m - (2j - 3)$
$= 6 - (2 \times 4 - 3) = 1$

Total degree of static indeterminacy, T = E + I = 2 + 1 = 3.

5. Find the kinematic indeterminacy of the truss shown in Fig. 3.

Ans. Two degrees of freedom are possible for each joint. There are five restraints.

∴ No. of possible independent joint displacements *i.e.*, kinematic indeterminacy,

K.I = 4 × 2 – 5 = 3.

6. Determine the degree of static indeterminacy of the frame shown in Fig. 4.

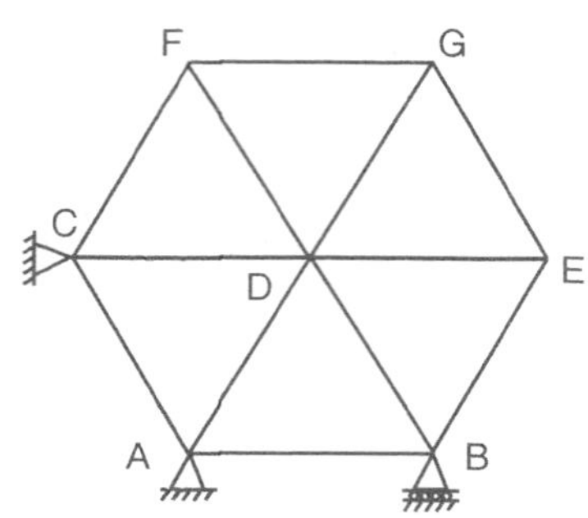

Fig. 4

Ans. Internal indeterminacy, $I = m - (2j - 3)$
$= 12 - (2 \times 7 - 3) = 12 - 11 = 1$

External indeterminacy, $E = r - R = 2 + 2 + 1 - 3 = 5 - 3 = 2$

Total degree of static indeterminacy, T = E + I = 2 + 1 = 3.

7. Find the degree of static indeterminacy of the truss shown in Fig. 5.

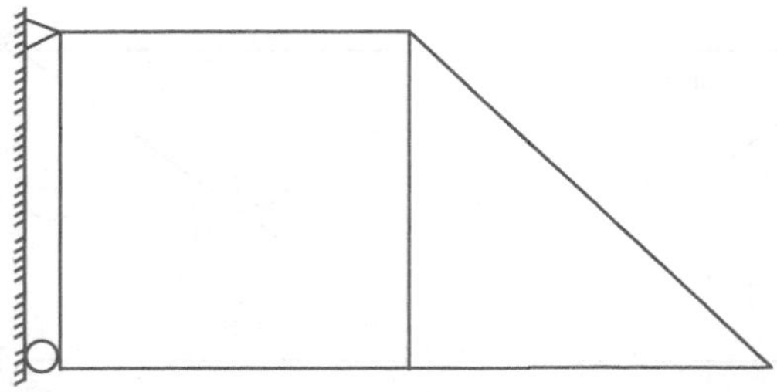

Fig. 5

Ans. The structure is externally determinate.

Internal indeterminacy, $I = m - (2j - 3)$

$= 6 - (2 \times 5 - 3) = 6 - 7 = -1$

Hence the truss is unstable (or deficient).

8. To analyse the truss shown in Fig. 6, member BD was removed and the truss made determinate. The forces in AD and DC were F_1 and F_3 due to the external force P. Subsequently a 1 kN force was applied along line DB and forces in AD and DC were k_1 and k_2. Assuming that the values of $\frac{AE}{l}$ is the same for all the 3 members, determine the force in the redundant member (BD) due to P.

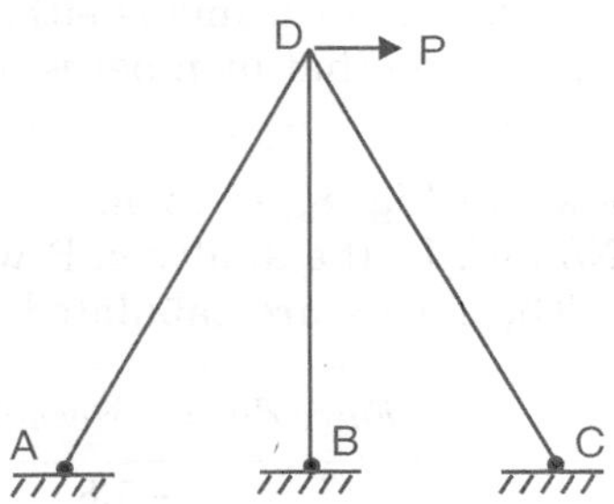

Fig. 6

Ans. $$F_{BD} = \frac{F_1 k_1 + F_2 k_2}{k_1^2 + k_2^2 + 1}$$

9. In the externally redundant frame shown in Fig. 7, the support at L_3 was removed. The remaining structure was analysed for the external loads W_1 and W_2. The resulting forces in the members were $F_1, F_2, \ldots\ldots, F_9$. Again, the remaining structure was analysed for an upward load of 1 kN at L_3. The resulting forces in the members were $k_1, k_2, \ldots\ldots, k_9$. Assuming that $\frac{AE}{l} = 10^4$ kN/m for all the members, find the reaction at L_3.

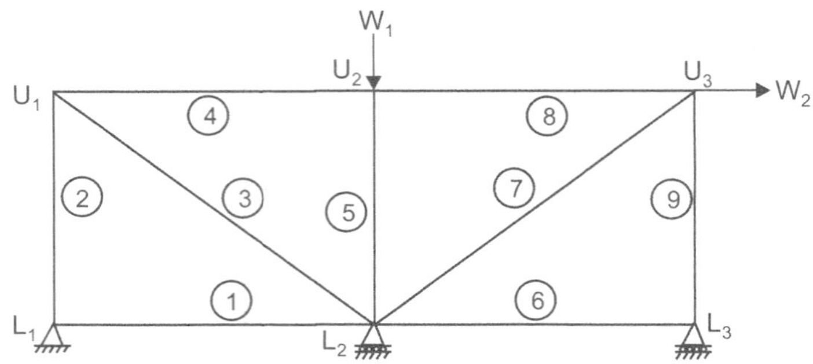

Fig. 7

Ans. $R = -\dfrac{\Sigma F_i k_i}{\Sigma k_i^2}$.

10. What are the methods available for the analysis of redundant frames ?

Ans. Methods available for the analysis of redundant frames are

1. Using Castigliano's theorem of minimum strain energy.
2. Clerk Maxwell's method (unit load method).
3. Matrix flexibility method.
4. Matrix stiffness method.

11. What is the effect of lack of fit in redundant frames ?

Ans. When any member of a redundant frame is either longer or shorter than what is required, it is said to have lack of fit. When that member is forced in position, stresses will be induced in all the members of the frame.

12. In the triangular truss shown in Fig. 8, all members have a length of 3.0 m, area of 600 mm^2 and E of 2000 kN/mm^2. In the analysis, B was kept on rollers and the truss with loads on C analysed. The forces are tabulated below :

Member	*Force due to external loads in kN*
1	– 1.8
2	– 18.2
3	+ 9.1

Find the horizontal reaction at B due to the given loading.

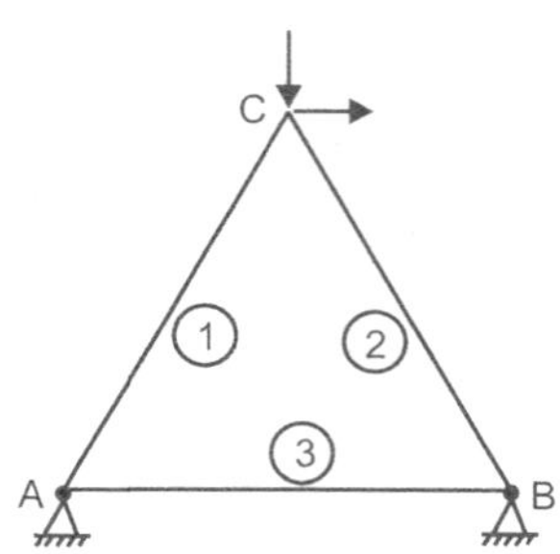

Fig. 8

Ans. The given forces can be called F_1, F_2 and F_3. If we apply a unit load at B along AB, we will get k_1, k_2, k_3. Obviously, $k_1 = k_2 = 0$ and $k_3 = 1$ (see Fig. 9).

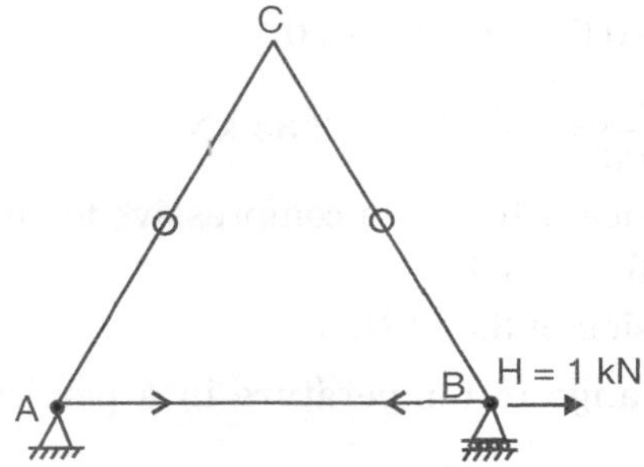

Fig. 9

The required redundant horizontal force

$$H = -\frac{\Sigma Fk \dfrac{l}{AE}}{\Sigma k^2 \dfrac{l}{AE}} = -\frac{(-1.8)\times 0 + (-18.2)\times 0 + 9.1\times 1}{1^2}$$

$$H = -9.1 \text{ kN}$$

Reaction H at B will be inwards (←).

13. In the frame ABCD shown in Fig. 10, a pair of unit forces were introduced at B and D. This produced member forces as given under.

Member	*Forces due to unit load*
AB	– 0.8
BC	– 0.6
CD	– 0.8
DA	– 0.6
AC	1.0

$\dfrac{AE}{l}$ is the same for all 5 members and is 2000 kN/m. A member BD identical to AC is now introduced in order to strengthen the truss. But its actual length was 5.005 m and it was forced in place. Find the force in member AC due to lack of fit.

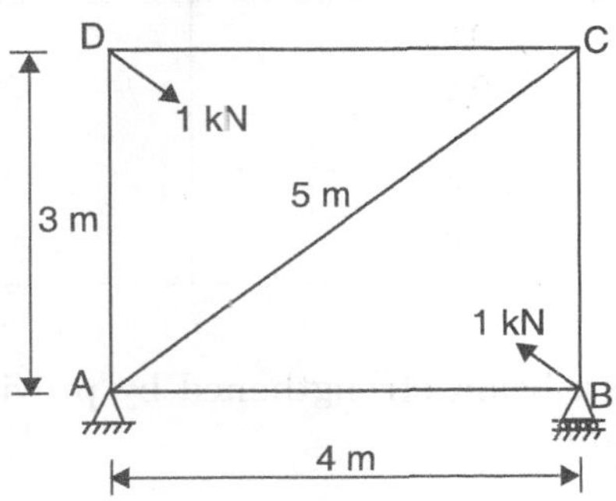

Fig. 10

Ans. The new member will develop a force X given by $X = \dfrac{-\delta}{\Sigma \dfrac{k^2 l}{AE}}$

$$\Sigma k^2 = 0.8^2 \times 2 + 0.6^2 \times 2 + 1^2 = 3.0$$

$$X = -\frac{0.005}{3.0/2000} = -\frac{5 \times 2}{3} = -3.33 \text{ kN}$$

DB which is forced in place will have a compressive force of 3.33 kN.

Force in AC = 1 × (– 3.33) = – 3.33.

AC will have a compression of 3.33 kN.

14. What is the effect of change in temperature in a particular member of a redundant frame ?

Ans. When any member of a redundant frame is subjected to a change in temperature, it will cause a change in length of that particular member, which in turn will cause lack of fit stresses in all other members of the redundant frame.

15. What is the difference in behaviour of the members of a statically determinate frame and of a redundant frame when a particular member of the frame is having lack of fit or subjected to change in temperature ?

Ans.

S.No.	*Details*	*Determinate frame*	*Redundant frame*
1.	Lack of fit of a member.	This will not affect the forces in members. In other words there is no question of lack of fit in determinate structures.	Lack of fit will cause forces in other members, when the member is forced in position.
2.	Change in temperature of a member	Will produce change in length of member and will not produce stresses in members.	Will produce change in length of member and hence will result in stresses in all members.

16. In question 13, find the forces in members due to the lack of fit of redundant member BD, taking force in BD as 3.33 kN compression.

Ans.

Member	*Force due to unit load, k*	*Forces due to X (= – 3.33) = k (– 3.33)*
AB	– 0.8	2.67 kN
BC	– 0.6	2.00 kN
CD	– 0.8	2.67 kN
DA	– 0.6	2.00 kN
AC	1.0	– 3.33 kN

17. What is a trussed beam ?

Ans. A trussed beam is a beam strengthened by providing ties and struts. This is normally adopted in timber beams.

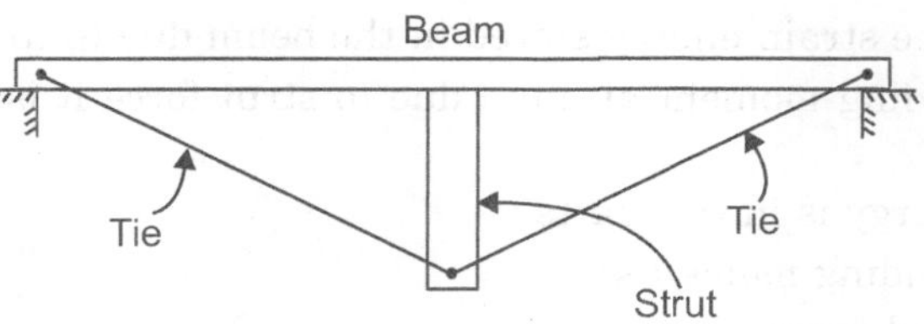

Fig. 11

18. What kind of strain energy is dominant in the various members of a trussed beam (Fig. 11) ?

Ans.

Member	*Strain energy due to*
Beam	Bending and axial force
Strut	Axial compression
Tie	Axial tension

19. Find the force in the tie of the trussed beam shown in Fig. 12 if the force in the strut is R.

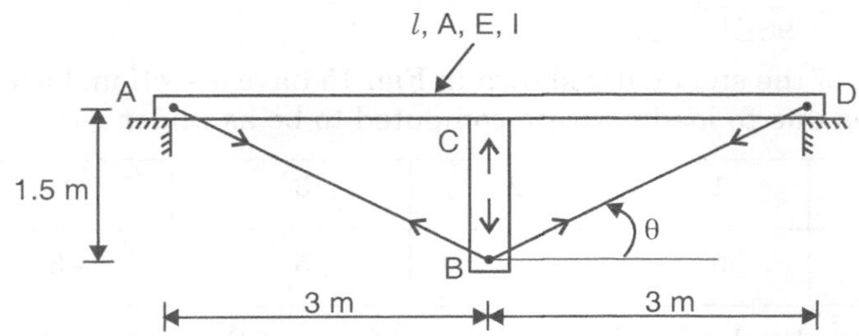

Fig. 12

Ans.

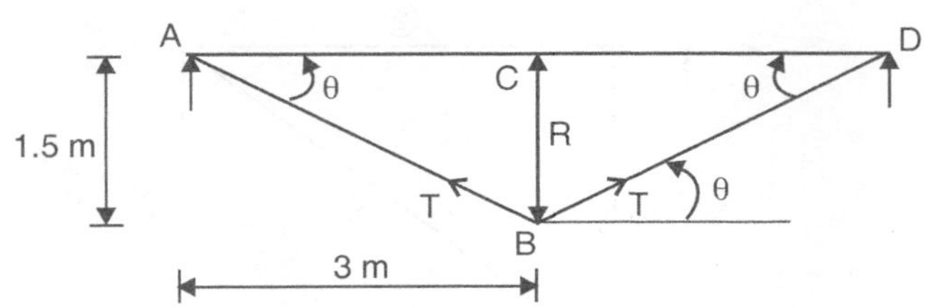

Fig. 13

Force in the strut BC = R

$$2T \sin \theta = R$$

$$\tan \theta = \frac{1.5}{3} = 0.5 \ ; \quad \theta = 26.57 \ ; \sin \theta = 0.447$$

$$T = \frac{1}{2 \times 0.447} R = 1.118 \text{ R.}$$

20. In Fig. 13, find the strain energy stored in the beam due to the force R on the strut.

Ans. (*a*) The bending moment at any x due to strut force R is (Fig. 14) M = Rx/2. The axial force = T cos θ = R

(*b*) The strain energy is in two parts :

(*i*) U_m (due to bending moments)

(*ii*) U_F (due to axial force)

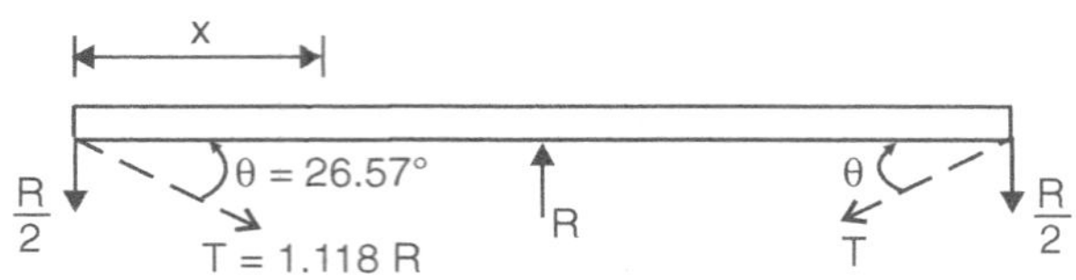

Fig. 14

$$U_m = 2\int_0^{l/2} \frac{M^2\,dx}{2EI} = 2\int_0^{l/2} \frac{R^2x^2}{4}\frac{dx}{2EI} = \frac{R^2}{4}\frac{l^3}{3\times 8}\frac{1}{EI} = \frac{R^2l^3}{96EI}$$

$$U_F = \frac{P^2l}{2AE} = \frac{R^2l}{2AE}$$

$$U = \frac{R^2l^3}{96EI} + \frac{R^2l}{2AE}$$

21. The members of the steel truss shown in Fig. 15 have a sectional area of 1000 mm². The internal forces due to loading were computed to be as under :

Member	1	2	3	4	5
Force in kN	0	– 4	5	– 3	0

What would be the changes in the internal forces if the cross-sectional area of member 3 is increased to 2000 mm² ?

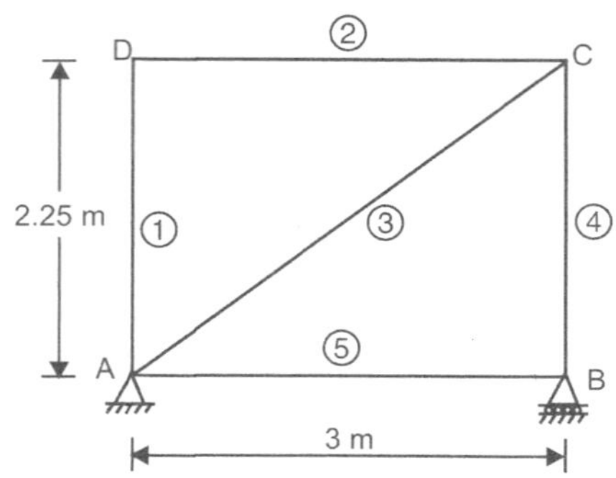

Fig. 15

Ans. There would be no change, since the truss is determinate. In a determinate truss the internal forces are independent of member cross-sections.

22. If instead of external force on the truss shown in Fig. 15, a unit force is introduced along BD at B and along DB at D, what would be the internal forces ?

Ans. The internal forces due to unit forces at B and D would be as under :

Member	1	2	3	4	5
Force in kN	– 0.6	– 0.8	1.0	– 0.6	– 0.8

23. For the truss in question 21, compute the strain energy stored in the truss due to external loading. (E = 2000 kN/mm^2)

Ans. Strain energy in a single member $= \dfrac{P^2 l}{2AE}$

Strain energy in the truss $= \sum_1^5 \dfrac{P^2 l}{2AE}$

$$U = \frac{1}{2AE}\,[0 + 4^2 \times 3000 + 5^2 \times 3750 + 3^2 \times 2250 + 0]$$

$$= \frac{1}{2 \times 1000 \times 2000}\,[48000 + 93750 + 20250]$$

$$U = 114 \text{ kNmm.}$$

24. For the truss in question 21, compute the strain energy in the 5 members due to a force of F along BD (instead of a unit force).

Ans. U_k = strain energy in the truss $= \sum_1^5 \dfrac{P^2 l}{2AE}$; $P = F.k$

$$U_k = \sum_1^5 \frac{F^2 k^2 l}{2AE}$$

$$U_k = \frac{F^2}{2AE}\,[0.36 \times 2250 + 0.64 \times 3000 + 1 \times 3750 + 0.36 \times 2250 + 0.64 \times 3000]$$

$$U_k = \frac{F^2}{2 \times 1000 \times 2000}\,[9210] = 2302.5 \times 10^{-6}\,F^2.$$

25. A cable stayed bridge has its deck DBE 12 m long supported by a pylon ABC and two cables CD and CE. Write an expression for bending moment at a section X on the deck when the compression in the pylon BC is 'F' kN (see Fig. 16).

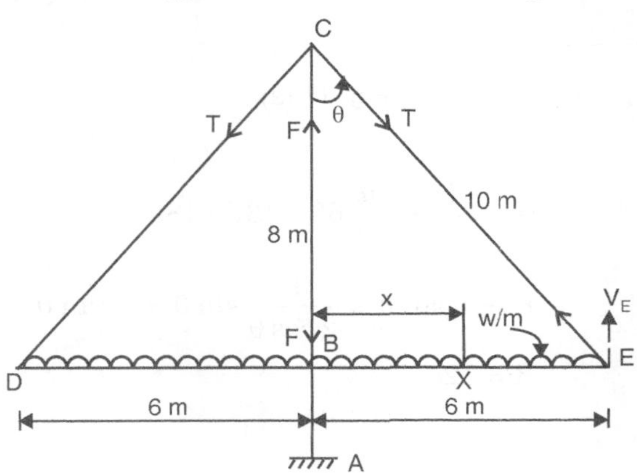

Fig. 16

Ans. From the vertical equilibrium of C

$$2T \cos \theta = F \ ; \qquad T = \frac{F}{2 \cos \theta}$$

At E, the vertical component of T is $V_E = T \cos \theta = \frac{F}{2}$

$$M_x = - w \frac{(6 - x)^2}{2} + V_E (6 - x)$$

$$M_x = \frac{F}{2} (6 - x) - \frac{w}{2} (6 - x)^2 .$$

26. In the bridge shown in Fig. 16, determine the force R in portion AB of the pylon due to the loading.

Ans. R = force in pylon = $w \times 12$.

27. For the bridge shown in Fig. 16, draw the free body of portion BE of the deck.

Ans.

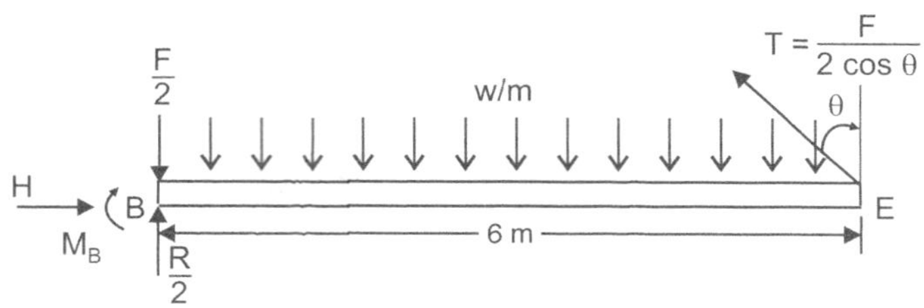

Fig. 17. Free body of BE.

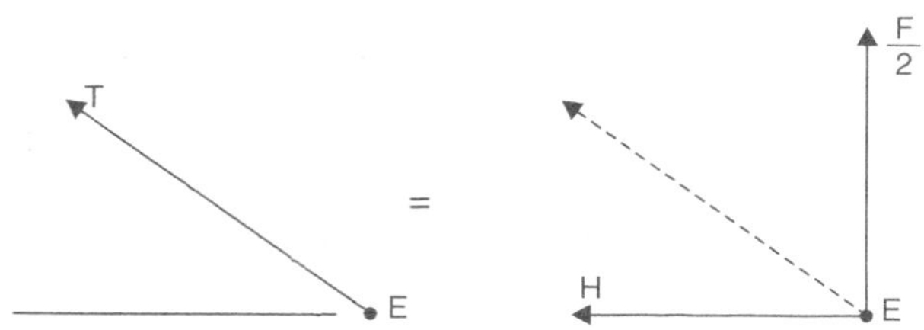

Fig. 17 (a)

$$M_B = \frac{F}{2} 6 - \frac{w}{2} 6^2 = (3F - 18w)$$

$$H = T \sin \theta = \frac{F}{2 \cos \theta} \sin \theta = \frac{F}{2} \tan \theta$$

$$R = 12w.$$

28.

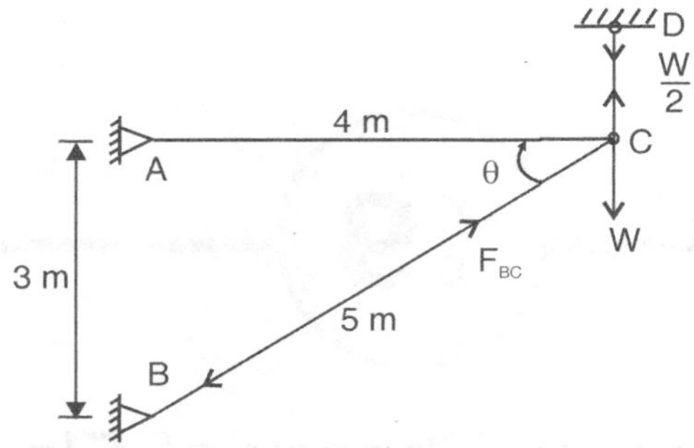

Fig. 18

The bracket ABC in Fig. 18 is strengthened by adding a tie CD.

(*i*) When CD takes half the external load, what are the forces in AC and BC ?

(*ii*) How can we redesign CD in order to reduce the force in BC further ?

Ans. (*i*) $\sin\theta = 0.6$; $\cos\theta = 0.8$

$$F_{BC}\sin\theta = \frac{W}{2}; \qquad F_{BC} = \frac{W}{2\sin\theta} = \frac{W}{1.2} \text{ (compressive)}$$

$$\text{Force in AC} = \frac{W}{1.2}\cos\theta$$

$$= \frac{W}{1.2}\times 0.8 = \frac{2}{3}\text{ W (tensile)}$$

(*ii*) To reduce F_{BC} further, increase the area of tie CD.

19

Introduction to Theory of Elasticity

1. What are 'surface forces' ?

Ans. Surface forces are the forces distributed/acting over the surface of a body.

Examples : Atmospheric pressure, hydrostatic pressure or contact pressure exerted by one body on another. Surface forces are expressed in terms of force per unit area.

2. What are 'body forces' ?

Ans. Body forces are the forces distributed over the volume of a body.

Examples : Gravitational force, electromagnetic force, centrifugal force and inertial forces. Body forces are expressed as force per unit volume.

3. Express the Cartesian stress components at a point, in three dimensions, in matrix form.

Ans. The state of stress at a point in Cartesian co-ordinates can be represented by the following nine components of stresses known as stress tensor.

$$[\sigma] = \begin{bmatrix} \sigma_x & \tau_{xy} & \tau_{xz} \\ \tau_{yx} & \sigma_y & \tau_{yz} \\ \tau_{zx} & \tau_{zy} & \sigma_z \end{bmatrix}$$

σ = Normal stress ; τ = Shear stress.

4. State the equations of equilibrium, in three dimensions, in Cartesian coordinates.

Ans. Equations of equilibrium :

$$\frac{\partial \sigma_x}{\partial x} + \frac{\partial \tau_{xy}}{\partial y} + \frac{\partial \tau_{xz}}{\partial z} + X = 0$$

$$\frac{\partial \tau_{yx}}{\partial x} + \frac{\partial \sigma_y}{\partial y} + \frac{\partial \tau_{yz}}{\partial z} + Y = 0$$

$$\frac{\partial \tau_{zx}}{\partial x} + \frac{\partial \tau_{zy}}{\partial y} + \frac{\partial \sigma_z}{\partial z} + Z = 0$$

X, Y and Z are body forces in the x, y and z directions respectively.

5. What are 'stress invariants' ?

Ans. Stress invariants are defined as the combinations of stresses at a point which remain the same for all variations of the orientation of the coordinate axes.

$$I_1 = \sigma_x + \sigma_y + \sigma_z = \text{Trace of the stress vector}$$
$$I_2 = \sigma_x\sigma_y + \sigma_y\sigma_z + \sigma_z\sigma_x - \tau^2_{xy} - \tau^2_{yz} - \tau^2_{zx}$$
$$I_3 = \text{determinant of the stress vector}$$

I_1, I_2 and I_3 are the first, second and third stress invariants respectively (which are combinations of given stresses).

6. What is 'Hydrostatic stress' ?

Ans. If $\sigma_x = \sigma_y = \sigma_z$, then $\sigma_1 = \sigma_2 = \sigma_3$ would also be equal.

This state of stress is called 'Hydrostatic'. For this state of stress, every plane is a principal plane and stresses in all directions are the same.

7. Give the equations of equilibrium in two dimensions in the absence of body forces (in Cartesian coordinates).

Ans. Equations of equilibrium :

$$\frac{\partial \sigma_x}{\partial x} + \frac{\partial \tau_{xy}}{\partial y} = 0$$

$$\frac{\partial \tau_{yx}}{\partial x} + \frac{\partial \sigma_y}{\partial y} = 0$$

8. The state of plane stress at a point is given by :

$\sigma_x = 75$ MPa ; $\sigma_y = -35$ MPa ; $\tau_{xy} = 20$ MPa

Determine the principal stresses.

Ans. Principal stresses are given by

$$\sigma_1 \text{ or } \sigma_2 = \frac{\sigma_x + \sigma_y}{2} \pm \sqrt{\left(\frac{\sigma_x - \sigma_y}{2}\right)^2 + \tau_{xy}^2}$$

$$= \frac{75 + (-35)}{2} \pm \sqrt{\left(\frac{75 - (-35)}{2}\right)^2 + 20^2}$$

$$= \frac{75 - 35}{2} \pm \sqrt{\left(\frac{75 + 35}{2}\right)^2 + 20^2} = 20 \pm 58.52$$

$\sigma_1 = 78.52$ MPa ; $\sigma_2 = -38.52$ MPa.

9. What are the co-ordinate systems available to represent the state of stress at a point?

Ans. The co-ordinate systems are:

(*i*) Cartesian co-ordinates (x, y, z)

(*ii*) Polar co-ordinates (r, θ)

(*iii*) Cylindrical co-ordinates (r, θ, z) and

(*iv*) Spherical co-ordinates (r, θ, ϕ).

10. The components of stress at a point are given by $\sigma_x = 30$ N/mm^2 ; $\sigma_y = 45$ N/mm^2; $\sigma_z = 40$ N/mm^2 ; $\tau_{xy} = 25$ N/mm^2; $\tau_{yz} = 15$ N/mm^2; $\tau_{zx} = 35$ N/mm^2. Determine the first two stress invariants.

Ans. First invariant of stress, $I_1 = \sigma_x + \sigma_y + \sigma_z$

$= 30 + 45 + 40 = 115.$

Second invariant of stress, $I_2 = \sigma_x\sigma_y + \sigma_y\sigma_z + \sigma_z\sigma_x - \tau_{xy}^2 - \tau_{yz}^2 - \tau_{zx}^2$

$= 30 \times 45 + 45 \times 40 + 40 \times 30 - (25)^2 - (15)^2 - (35)^2$

$= 2275$

$\therefore \quad I_1 = 115 \; ; I_2 = 2275.$

11. Give the characteristic (cubic) equation for strain at a point, in three dimensions, in cartesian co-ordinates.

Ans. The characteristic equation for strain at a point in three dimensions is given by,

$$\varepsilon^3 - J_1 \varepsilon^2 + J_2 \varepsilon - J_3 = 0$$

where J_1, J_2 and J_3 are the first, second and third invariants of strain respectively.

12. The stress invariants for the three dimensional state of stress at a point are : $I_1 = 120$; $I_2 = 1800$ and $I_3 = -25000$. Write down the cubic equation for the stresses.

Ans. The cubic equation is given by,

$$\sigma^3 - I_1 \sigma^2 + I_2 \sigma - I_3 = 0$$

Substituting, I_1, I_2 and I_3, the equation becomes

$$\sigma^3 - 120 \sigma^2 + 1800 \sigma - (-25000) = 0$$

i.e., $$\sigma^3 - 120 \sigma^2 + 1800 \sigma + 25000 = 0$$

The 3 roots of this equation are the principal stresses. The roots are,–8, 29 and 99.

13. The displacements at a point in a solid body are given by $u = a_1x$; $v = a_2y$; $w = a_3z$. Find the components of strain.

Ans.

$$\varepsilon_{xx} = \frac{\partial u}{\partial x} = \frac{\partial}{\partial x}(a_1x) = a_1$$

$$\varepsilon_{yy} = \frac{\partial v}{\partial y} = \frac{\partial}{\partial y}(a_2y) = a_2$$

$$\varepsilon_{zz} = \frac{\partial w}{\partial z} = \frac{\partial}{\partial z}(a_3z) = a_3$$

$$\gamma_{xy} = \left(\frac{\partial u}{\partial y} + \frac{\partial v}{\partial x}\right) = (0 + 0) = 0$$

$$\gamma_{yz} = \left(\frac{\partial v}{\partial z} + \frac{\partial w}{\partial y}\right) = (0 + 0) = 0$$

$$\gamma_{zx} = \left(\frac{\partial w}{\partial x} + \frac{\partial u}{\partial z}\right) = (0 + 0) = 0$$

14. The components of strain in a body are given by the strain matrix,

$$[\varepsilon] = \begin{bmatrix} 0 & 0 & -0.15 \\ 0 & 0 & 0.20 \\ -0.15 & 0.20 & 0 \end{bmatrix}$$

Find the characteristic equation given the principal strains as its roots.

Ans. The characteristic equation for strain at a point in three dimensions is given by,

$$\varepsilon^3 - J_1 \varepsilon^2 + J_2 \varepsilon - J_3 = 0$$

$$J_1 = \varepsilon_x + \varepsilon_y + \varepsilon_z = 0 + 0 + 0 = 0$$

$$J_2 = \varepsilon_x\varepsilon_y + \varepsilon_y\varepsilon_z + \varepsilon_z\varepsilon_x - \frac{\gamma_{xy}^2}{4} - \frac{\gamma_{yz}^2}{4} - \frac{\gamma_{zx}^2}{4}$$

$$= 0 + 0 + 0 - 0 - (0.2)^2 - (-0.15)^2$$

$$= -0.04 - 0.0225 = -0.0625$$

$$J_3 = 0$$

$$\therefore \quad \varepsilon^3 - 0 + (-0.0625)\,\varepsilon - 0 = 0$$

$\varepsilon^3 - 0.0625\varepsilon = 0$ is the characteristic equation. $\varepsilon_1 = 0$; $\varepsilon_2 = -0.25$; $\varepsilon_3 = 0.25$.

15. What are 'compatibility equations' ?

Ans. The equations resulting from the application of the strain equations are known as the compatibility equations.

The strains at a point in different co-ordinate directions have to be *consistent* with a *displacement field.* Thus the six strain components shall be *compatible.* So also the stress components. The equations resulting from the consistency conditions are *compatibility equations.*

16. Give the differential equation of compatibility in two dimensions in terms of (*i*) strains and (*ii*) stresses.

Ans. Compatibility equation in two dimensions (cartesian co-ordinates) are

$$\frac{\partial^2 \varepsilon_x}{\partial y^2} + \frac{\partial^2 \varepsilon_y}{\partial x^2} = \frac{\partial^2 \gamma_{xy}}{\partial y \partial x} \qquad \text{(in terms of strain)}$$

$$\left(\frac{\partial^2}{\partial x^2} + \frac{\partial^2}{\partial y^2}\right)(\sigma_x + \sigma_y) = 0 \qquad \text{(in terms of stress)}$$

17. What are plane stress problems ? Give an example.

Ans.

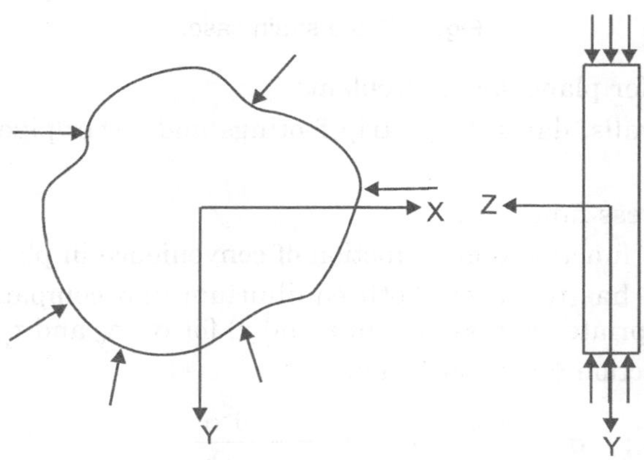

Fig. 1. Thin plate with in-plane loadin–an example for plane stress problem.

If the state of stress in a body can be specified by σ_x, σ_y and τ_{xy} only, then it is a plane stress problem. The other stresses namely σ_z, τ_{xz} and τ_{yz} would be zero.

Consider a thin plate loaded by forces applied at the boundary, parallel to the plane of the plate and distributed uniformly over the thickness. The stress components σ_z, τ_{xz}, τ_{yz} are zero on both faces of the plate. They may be assumed to be zero within the plate also. The stresses are then functions of x and y only and independent of z. Hence it is a plane stress problem.

18. What are plane strain problems ?

Ans. Problems involving a long body whose geometry and loading do not vary longitudinally are referred to as plane strain problems.

Here $\varepsilon_z = 0$ (since $w = 0$; conditions are the same at all cross-sections)

Stresses can be expressed in terms of σ_x, σ_y and τ_{xy} only.

ε_x, ε_y, would be γ_{xy} present.

$$\varepsilon_z = 0$$

$$\gamma_{xz} = \gamma_{yx} = 0$$

σ_x, σ_y and σ_z would be present.

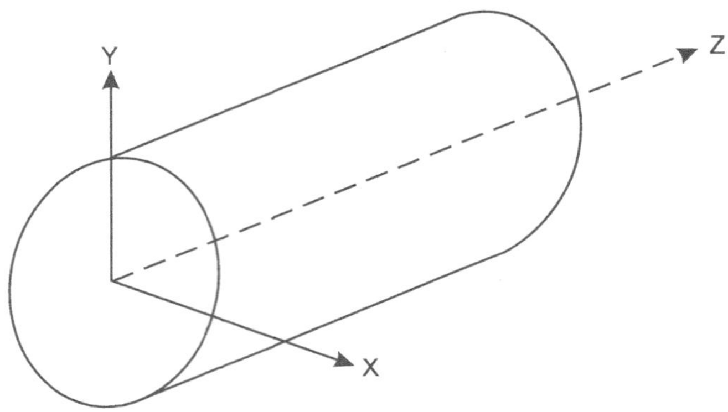

Fig. 2. Plane strain case.

19. Give 4 examples for plane strain problems.

Ans. Retaining walls, dams, long strip footings and sewer/pipe lines are examples of plane strain problems.

20. What is Airy's stress function ?

Ans. Airy's stress function ϕ is a function of convenience in plane stress problems.

A stress problem has to satisfy both equilibrium and compatibility conditions. The solution will seek appropriate expressions (in x and y) for σ_x, σ_y and τ_{xy}.

If we define a function $\phi(x, y)$ such that

$$\sigma_x = \frac{\partial^2 \phi}{\partial y^2} \;; \quad \sigma_y = \frac{\partial^2 \phi}{\partial x^2} \quad \text{and} \quad \tau_{xy} = -\frac{\partial^2 \phi}{\partial x \partial y}$$

then the equilibrium conditions connecting σ_x, σ_y and τ_{xy} will automatically get satisfied.

The compatibility condition will reduce to

$$\left(\frac{\partial^2}{\partial x^2} + \frac{\partial^2}{\partial y^2}\right)\left(\frac{\partial^2 \phi}{\partial x^2} + \frac{\partial^2 \phi}{\partial y^2}\right) = 0$$

Now, we only have to find ϕ to fit into the above 4th order partial differential equation. That will give the complete solution to the stress field problem.

20

Introduction to the Finite Element Method

1. What is finite element method ?

Ans. Finite element method is a very popular numerical technique for the computer solution of complex problems in engineering. Here, the principle of discretisation is used and the structure is considered to be an assemblage of discrete elements. This method is applicable to almost all problems wherever a variational formulation of the physical phenomenon is feasible.

2. What is a continuum ?

Ans. A continuum is defined as a domain in which matter exists at every point. That is, a continuum consists of an infinite number of particles and have a continuous surface bulk.

3. What is the basic principle involved in the finite element method ?

Ans. In the finite element method, the principle of discretisation is used. Here, the physical body is approximated by replacing it with an assemblage of discrete elements called 'finite elements'. For any typical element an approximate numerical model relating the displacement at its nodes and the displacement field over the body of the element is developed. When the elements are 'assembled' with due regard to equilibrium/compatibility, the displacement field over the entire assemblage and hence the stress field over the assemblage is obtained.

4. What are the specialties of finite element method ?

Ans. Finite element method permits the incorporation of complexities such as non-homogenous media, non-linear material behaviour, dynamic loading, *in situ* stress conditions, arbitrary geometrics, discontinuities and other factors that cannot be readily handled by the closed form and other classical procedures.

5. Why is the displacement based finite element method more widely used ?

Ans. Displacement based finite element method is more widely used because it is simple and easily applicable to different types of problems.

6. What is meant by discretisation ?

Ans. Discretisation is the process of subdividing a given body into a number of elements which results in an equivalent body of finite elements.

7. What are the factors that govern the selection of the type of finite element ?

Ans. Selection of type of finite element depends upon :

(*i*) the geometry of the body (1D, 2D or 3D) or structure and

(*ii*) the number of independent space co-ordinates *i.e.*, x, y, z necessary to describe the problem.

(*iii*) the nature of stress variations expected (whether bending or torsional stresses are expected, if contact stresses are possible, etc.)

8. Define 'nodes' and 'nodal lines'.

Ans. Nodes are points at which different elements are connected to generate the 'structure' or 'system'. Each node will have finite displacements.

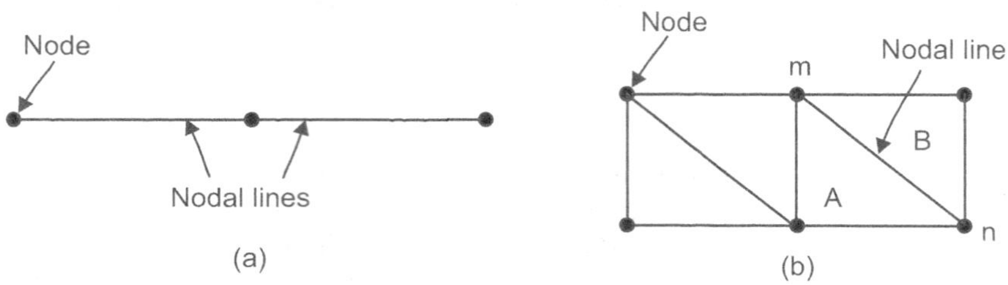

Fig. 1

If 2 nodes (Fig. 1) m and n are connected, the displacements of m as part of element B are the same as the displacements of m as part of element A. The same is true of node n. Since we adopt the same form of displacement model for A and B, contact at m and n will ensure contact between A and B all along the nodal line mn.

9. What are linear elements and higher order elements ?

Ans. Finite elements with straight sides are known as linear elements and those with curved sides are called higher order elements.

10. What is plane stress ?

Ans. Plane stress is defined to be a state of stress in which the normal stress and the shear stresses directed perpendicular to the plane are assumed to be zero.

11. What is plane strain ?

Ans. Plane strain is defined to be a state of strain in which the strain normal to the x-y plane ε_z and the shear strains ν_{xz} and ν_{yz} are assumed to be zero.

12. Give the relation between stresses and strains in a plane stress problem.

Ans. In a plane stress problem

$$\begin{Bmatrix} \sigma_x \\ \sigma_y \\ \tau_{xy} \end{Bmatrix} = \frac{E}{1-\nu^2} \begin{bmatrix} 1 & \nu & 0 \\ \nu & 1 & 0 \\ 0 & 0 & \dfrac{1-\nu}{2} \end{bmatrix} \begin{Bmatrix} \varepsilon_x \\ \varepsilon_y \\ r_{xy} \end{Bmatrix}$$

$$\{\sigma\} = [D]\,\{\varepsilon\}$$

13. Why is a constant strain triangular element called so ?

Ans. Constant strain triangular element (CST element) is called so because it has a constant strains throughout its area. ε_x, ε_y and r_{xy} values do not vary within the element.

14. What are the types of finite elements used for plane stress/strain problems ?

Ans. For plane stress/strain problems, quadrilateral, rectangular or triangular elements are used. Within the element, constant strain, constant stress or linearly varying stress/strain, etc. can be adopted.

15. What are the most obvious locations for nodes or nodal lines in the division of a structure/continuum into finite elements ?

Ans. Most obvious locations for nodes or nodal lines are,

(*i*) points of application of concentrated loads

(*ii*) locations where there are changes in intensity of loading

(*iii*) locations where there are discontinuities in geometry of the structure

(*iv*) interfaces between materials of different properties.

16. What is meant by 'Element Aspect Ratio' ?

Ans. The 'Element Aspect Ratio' is defined as the ratio of the largest dimension of the element to the smallest dimension.

17. What is the significance of 'Aspect ratio' ?

Ans. An optimum aspect ratio has to be chosen for better quality of the solution. The selection depends upon the difference in rate of change of displacements in different directions. If the displacements vary at about the same rate in each direction, the closer the aspect ratio to unity, the better will be the quality of the solution.

18. What is 'semiband width' in a stiffness matrix ?

Ans. When the set of equations using the finite element method are listed, it would show that all non zero values fall between two band lines which can be constructed parallel to the leading diagonal. The distance expressed as number of columns from the diagonal to the band line is called the semiband width (half band width). In the matrix shown below, the half band width is 2.

Half band width

$$\begin{bmatrix} 5 & 2 & 0 & 0 & 0 & 0 \\ 2 & 3 & 4 & 0 & 0 & 0 \\ 0 & 4 & 6 & 7 & 0 & 0 \\ 0 & 0 & 7 & 8 & 1 & 0 \\ 0 & 0 & 0 & 1 & 2 & 0 \\ 0 & 0 & 0 & 0 & 0 & 9 \end{bmatrix}$$

19. Give the relation between node numbering, degrees of freedom and half band width.

Ans. The relation is, $b = (d + 1)f$

where b is the half band width d is the largest difference between the system node numbers in a single assembled element and f is the degree of freedom at each node.

20. Two different systems of labelling the nodes for a body are shown in Fig. 2 (*a*) and (*b*). Which is better and why ? Degree of freedom at each node may be taken as 2.

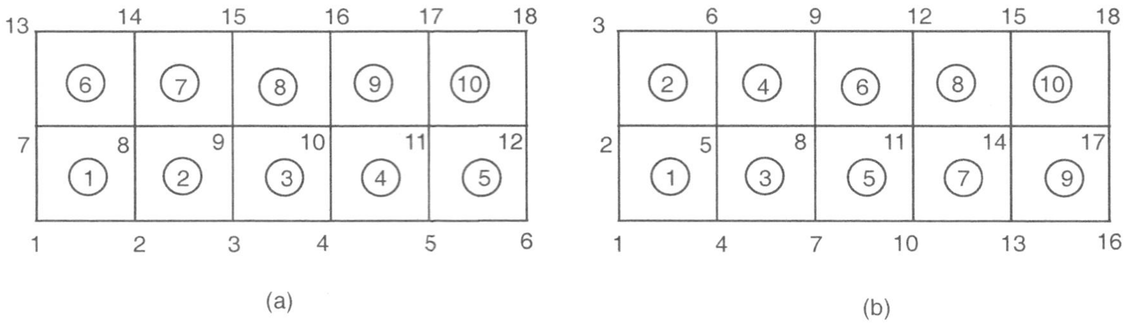

Fig. 2

Ans. $f = 2$; Degree of freedom for 18 nodes, $n = 18 \times 2 = 36$

Details	*(a) d = 7*	*(b) d= 4*
Semiband width $b = (d + 1)f$	$b = (7 + 1) \times 2 = 16$	$b = (4 + 1) \times 2 = 10$
d, o.f. of the system × half band width $(n \times b)$	$36 \times 16 = 576$	$36 \times 10 = 360$
Solution time	More	Less

∴ The labelling system given in (*b*) is better since it saves the computer storage space and solution time.

21. What is meant by 'displacement models' ?

Ans. In finite element analysis, after discretising the body into a number of finite elements, simple functions are assumed to approximate the displacements of each element. These functions are called 'displacement models' or displacement functions or displacement fields. The unknown quantities of the displacement functions are the displacements at the nodal points.

22. What is meant by 'Pascal Triangle' ? How is it used in choosing displacement models ?

Ans. Pascal triangle is a triangular array shown below representing the terms in a complete two-dimensional polynomial.

When we choose a displacement model of an element in a polynomial form, we better be guided by the Pascal triangle. Thus a linear model will contain 3 terms, the x, y and constant terms. A 2 degree model shall include the top 3 rows. We cannot omit row one or two and include row 3. A 3 degree model shall include rows 1, 2 and 3 before including any term from row 4.

$$
\begin{array}{ccccccccc}
 & & & & 1 & & & & \\
 & & & x & & y & & & \\
 & & x^2 & & xy & & y^2 & & \\
 & x^3 & & x^2y & & xy^2 & & y^3 & \\
x^4 & & x^3y & & x^2y^2 & & xy^3 & & y^4
\end{array}
$$

23. What is meant by convergence ?

Ans. In a finite element model, the results obtained from a coarse grid are likely to be in error compared to the correct result (which in most cases we would not know). When we make the mesh progressively finer, the results shall approach some quantity and not keep fluctuating.

A model in which the error goes on getting smaller and smaller as we use finer and finer grids is supposed to *converge* to the correct solution.

The choice of the displacement model has a major influence on the ability of the finite element model to converge (and not diverge or oscillate).

Fig. 3 explains the concept of convergence.

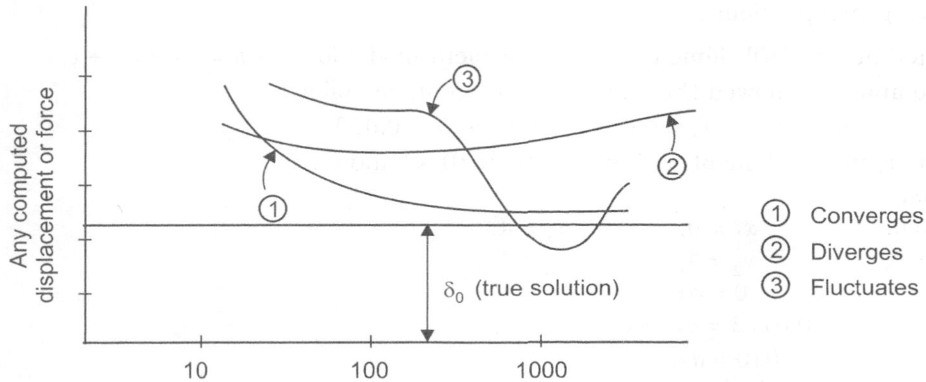

Fig. 3. No. of elements in the model.

24. What are the features of a good assembled stiffness matrix ?

Ans. Features of a stiffness matrix are,

(*i*) The system matrix will have its non zero terms clustered about its main diagonal while locations distant from the diagonal are zero terms. The matrix will be banded.

(*ii*) The elements of a system matrix are symmetric $k_{ij} = k_{ji}$

(*iii*) Diagonal terms are positive definite.

25. What are the methods available for the solution of a positive definite, symmetric and banded matrix?

Ans. Solution methods are,

(*i*) Cholesky's square root method

(*ii*) Half band Gauss elimination solution technique

(*iii*) Half block technique

(*iv*) Skyline technique and

(*v*) Frontal solution.

26. What are the primary unknowns in a displacement based finite element analysis ?

Ans. In a displacement based finite element analysis, the nodal displacements will be obtained first and they are the primary unknowns.

27. What are secondary unknowns in a displacement based finite element analysis ?

Ans. From the primary unknowns, namely the displacements of nodes, strains will be obtained. Using the stress-strain relationships, the stresses and thereby the principal stresses will be obtained. These stresses are in fact the secondary unknowns.

28. For a 3 noded 6 degree of freedom triangular element the following displacement models are proposed.

$$u = ax^2 + bx + c$$
$$v = dy^2 + ey + f$$

Comment on their validity.

$$1$$
$$x \quad y$$
$$x^2 \quad xy \quad y^2$$
$$x^3 \quad x^2y \quad xy^2 \quad y^3$$
$$x^4 \quad x^3y \quad x^2y^2 \quad xy^3 \quad y^4$$

Ans. These are not good models since they skip some terms in the Pascal triangle. The given models for u and v skip a part of row 2 and go to row 3. This model will have convergence problems.

29. In a 3 noded CST element the displacement model for u is $u = ax + by + c$. The analysis showed the nodal displacements as follows :

$$u_1 = 0, \quad u_2 = 0.0172 \quad \text{and} \quad u_3 = -0.010$$

Find ε_x in the element with nodes (0, 0), (0, 4) and (4, 4).

Ans.

$x_1 = 0,\quad x_2 = 0,\quad x_3 = 4,$

$y_1 = 0,\quad y_2 = 4,\quad y_3 = 4$

$$0 = ax_1 + by_1 + c \quad ...(1)$$
$$0.0172 = ax_2 + by_2 + c \quad ...(2)$$
$$-0.010 = ax_3 + by_3 + c \quad ...(3)$$

Applying the values of x_i and y_i

(1) $0 = 0 + 0 + c \quad \therefore \quad c = 0$

(2) $0.0172 = 0 + b \times 4 \quad\quad b = 0.0043$

(3) $-0.010 = a \times 4 + b \times 4$

$$4a = -0.010 - 0.0043 \times 4$$
$$a = -0.0025 - 0.0043$$
$$a = -0.0068$$

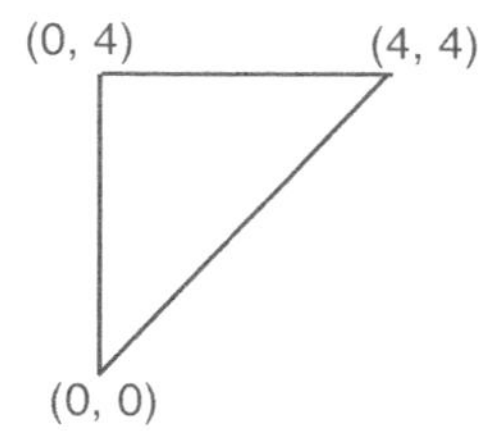

Fig. 4

$$\varepsilon_x = \frac{\partial u}{\partial x} = a$$

$\varepsilon_x = -0.0068$ for the entire element.

30. A rectangular plate is divided into 4 elements as shown in Fig. 5. The values of u_i and v_i, the nodal displacements of the system got from the analysis are indicated in the figure. Find the nodal displacements of elements (3) and (2).

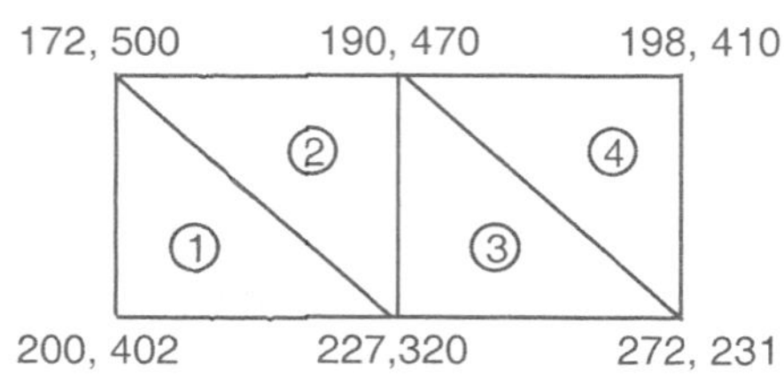

Fig. 5

Ans. Fig. 6 shows the nodal displacements of element (2) and element (3).

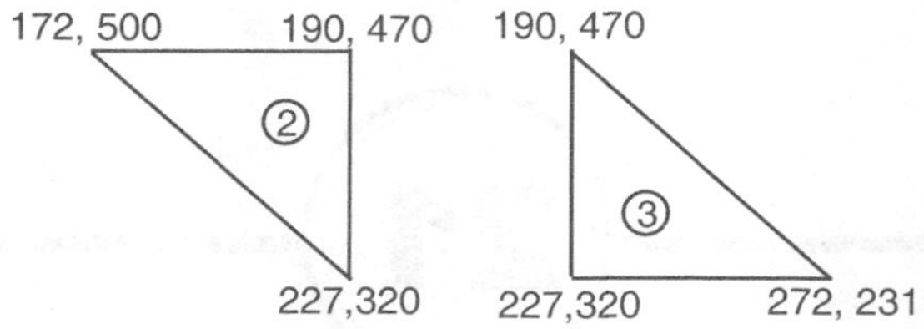

Fig. 6

31. Number the nodes of the system shown in Fig. 7 so as to get a system stiffness matrix of minimum band width.

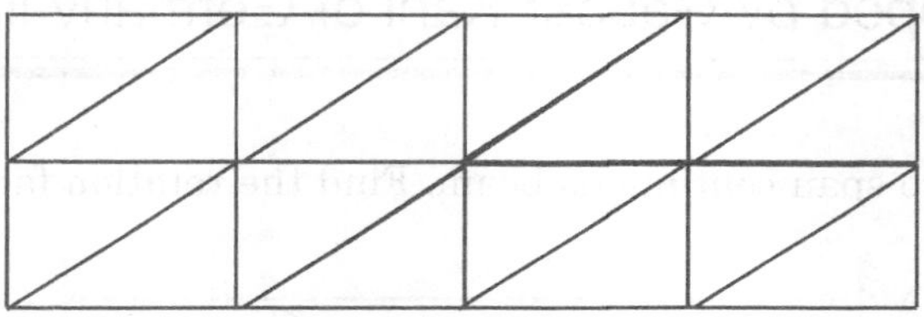

Fig. 7

Ans. Fig. 8 shows the node numbering that will yield minimum band width in the assembled stiffness matrix.

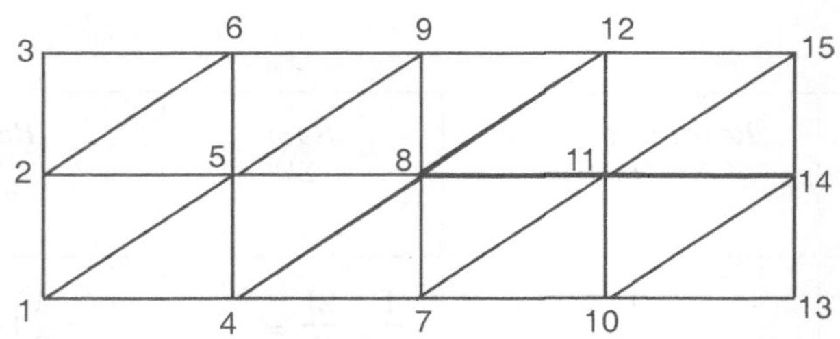

Fig. 8

21

Kani's Method

(Developed by Gasper Kani of Germany in 1947)

1. Fig. 1 shows a two span continuous beam. Find the rotation factors for the joint B.

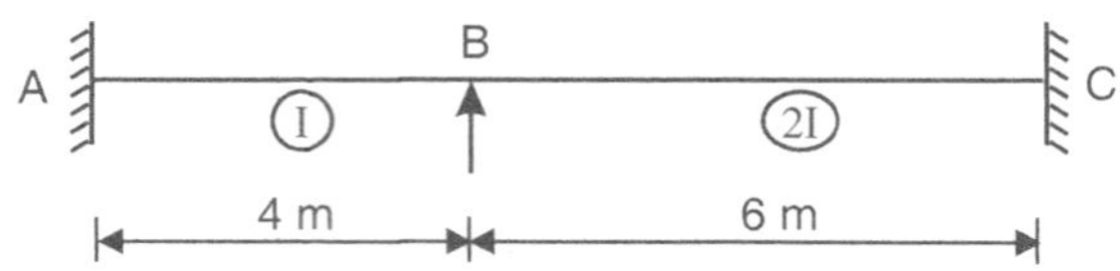

Fig. 1

Ans.

Joint	*Members*	*Relative stiffness* (*k*)	*Sum* Σk	*Rotation factor* $u = \left(-\frac{1}{2}\right)\frac{k}{\Sigma k}$
B	BA	$\frac{I}{4}$	$\frac{I}{4} + \frac{2I}{6} =$	$\left(-\frac{1}{2}\right) \times \frac{I}{4} \times \frac{12}{7I} = -0.214$
	BC	$\frac{2I}{6}$	$\frac{3I + 4I}{12} = \left(\frac{7I}{12}\right)$	$\left(-\frac{1}{2}\right) \times \frac{2I}{6} \times \frac{12}{7I} = -0.286$

2. Fig. 2 shows a two span continuous beam. Find the rotation factors for the joint B.

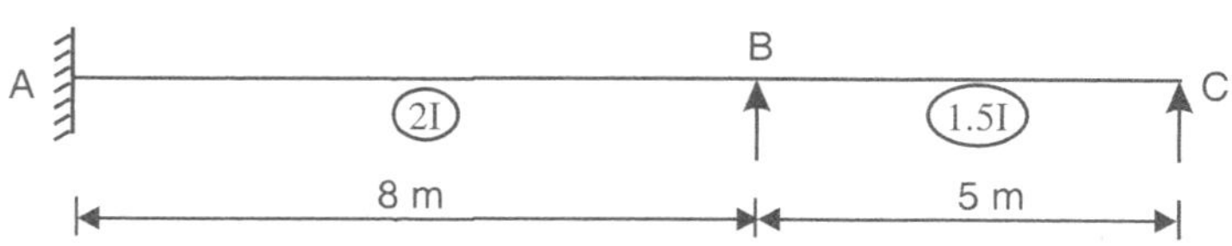

Fig. 2

Ans.

Joint	*Members*	*Relative stiffness* (k)	*Sum* Σk	*Rotation factor* $u = \left(-\frac{1}{2}\right)\frac{k}{\Sigma k}$
B	BA	$\frac{2I}{8} = \frac{I}{4}$	$\frac{I}{4} + \frac{4.5I}{20} =$	$\left(-\frac{1}{2}\right) \times \frac{I}{4} \times \frac{20}{9.5I} = -0.263$
	BC	$\frac{3}{4} \times \frac{1.5I}{5}$	$\frac{5I + 4.5I}{20} = \left(\frac{9.5I}{20}\right)$	$\left(-\frac{1}{2}\right) \times \frac{4.5I}{20} \times \frac{20}{9.5I} = -0.237$

3. Three members meet at a joint O as shown in Fig. 3. The rotation factor for the member OA and OB are – 0.22 and – 0.17 respectively. What is the rotation factor for the member OC ?

Ans. Sum of rotation factors of the members at a joint = – 0.5

Hence the rotation factor of member, OC

OC = – 0.5 – (– 0.22 – 0.17) = – 0.11.

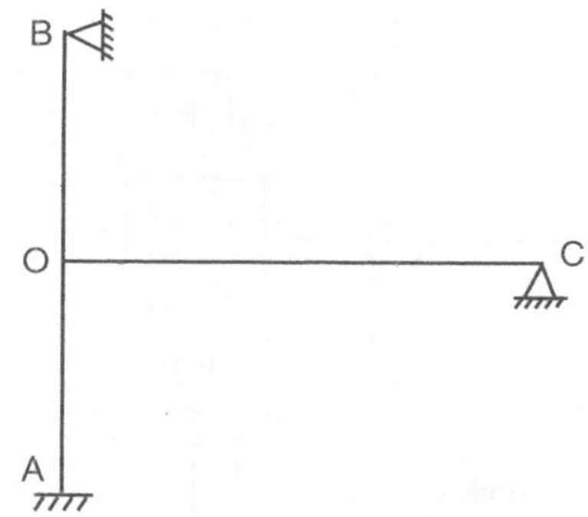

Fig. 3

4. What is the value of the rotation moment at a fixed end, considered in Kani's method ?

Ans. Rotation moment is the moment causing a rotation θ at the end A under consideration when the far end B is fixed, inducing half of the above moment at the far end. When end A is fixed, the rotation being zero, the rotation moment is also zero.

5. What is the relative stiffness adopted in Kani's method, when the far end of a member is hinged ?

Ans. When the far end B of a member AB is hinged or pinned, it is conveniently considered as fixed and the relative stiffness is taken as $\left(\frac{3}{4}\right)\frac{I}{l}$. Once this is done, no further rotation moment is assigned to B although B would rotate.

6. What is the relation between the fixed end moment and rotation moment of the ends of a member, in Kani's method ?

Ans.

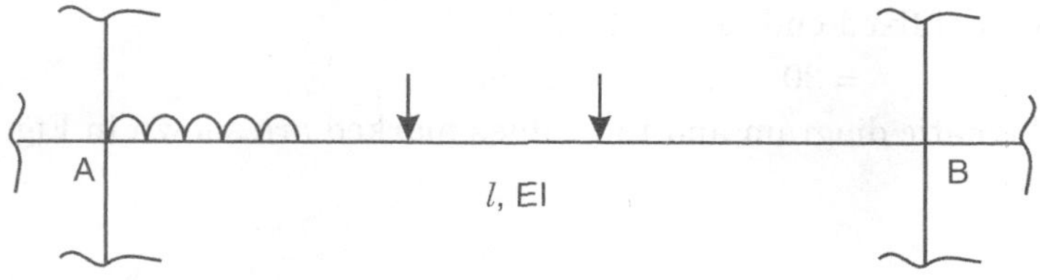

Fig. 4

Let us take, for example, a member AB of a frame in Fig. 4. If the final moment at A is M_{AB},

$$M_{AB} = M_{FAB} + 2M'_{AB} + M'_{BA}$$

where, M_{FAB} is the fixed end moment at A due to the applied loading

M'_{AB} is the rotation moment at A and

M'_{BA} is the carry over from B due to the rotation moment at B.

7. Sketch a schematic diagram for proceeding with the method of rotation contributions in Kani's method, for a typical joint.

Ans. The schematic diagram for a typical joint B is shown in Fig. 5.

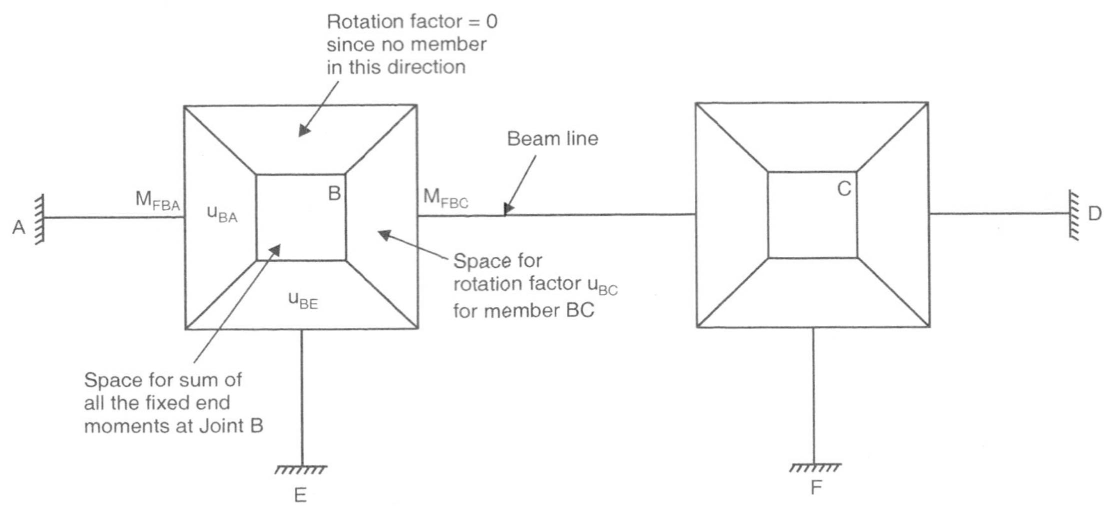

Fig. 5

8. In the portal frame shown in Fig. 6, the fixed end moments at joint B due to external loading are,

$$M_{FBA} = +30 \text{ kNm}$$
$$M_{FBC} = -50 \text{ kNm}$$

Rotation factors are,

$$u_{BA} = -0.20$$
$$u_{BC} = -0.30$$

Sketch the schematic diagram for proceeding with Kani's method and mark the available values at B.

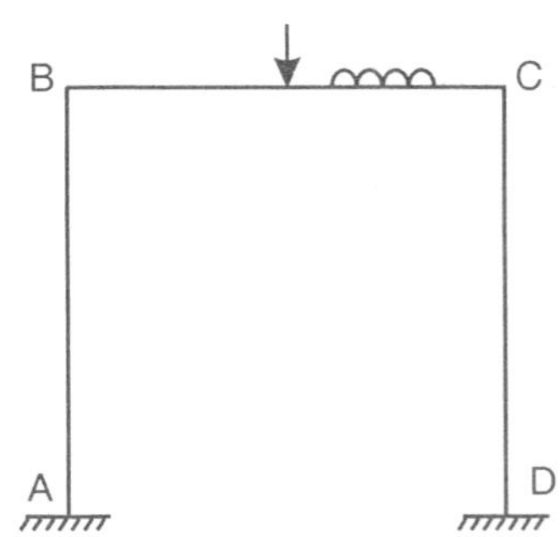

Fig. 6

Ans. Sum of fixed end moments at B

$$= 30 - 50 = -20 \text{ kNm}.$$

The schematic diagram and the values marked are shown in Fig. 6 (*a*).

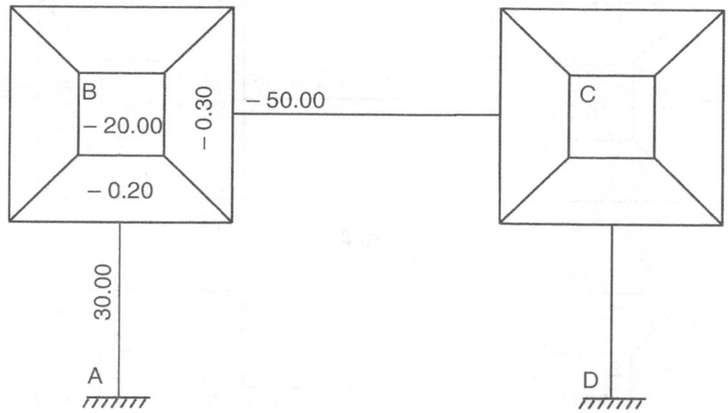

Fig. 6 (a)

9. Write down the equilibrium equation of a typical joint J in Kani's method (Fig. 7).

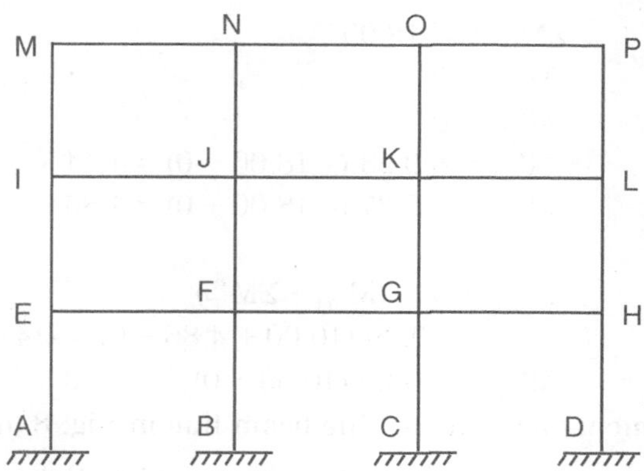

Fig. 7

Ans. For equilibrium of joint J, the sum of end moments at J must be zero.

$$\Sigma M_J = 0$$

i.e., $$\Sigma M_{FJ} + 2\,\Sigma M'_{JJ'} + \Sigma M'_{JJ} = 0$$

where, J′ represents the four nodes at the far ends namely I, N, K and F.

M_{FJ} are fixed end moments at J.

$M'_{JJ'}$, $M'_{J'J}$ are rotation moments at J.

10. For a three span continuous beam, the values of fixed end moments due to some loading and the rotation factors are marked shown in Fig. 8. Carry out the first cycle of iteration.

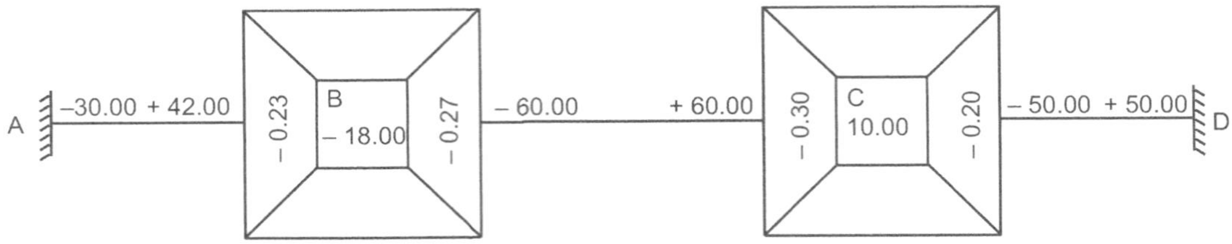

Fig. 8

Ans.

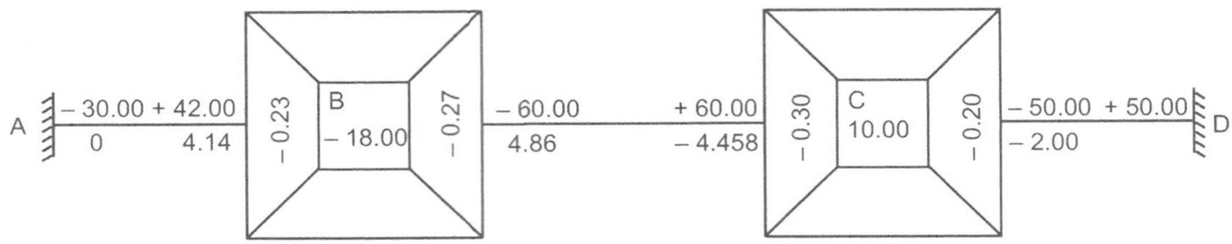

Fig. 8. (a) First cycle of iteration.

I cycle of iteration :

$$\Sigma M_{FB} = -18.00$$

$$M'_{AB} = 0 \qquad (\because \text{ A is fixed})$$

$$M'_{CB} = 0 \qquad (\text{Initially assumed})$$

$\therefore$

$$M'_{BA} = -0.23\,(-18.00 + 0) = 4.14$$

$$M'_{BC} = -0.27\,(-18.00 + 0) = 4.86$$

Joint C :

$$M'_{CB} = u_{CB}\,(M_{FC} + \Sigma M'_{C})$$

$$= -0.30\,(10.00 + 4.86 + 0) = -4.458$$

$$M'_{CD} = -0.20\,(10.00 + 0) = -2.00$$

These rotation moments are entered the beam line in Fig. 8 (*a*).

11. A three span continuous beam is loaded as shown in Fig. 9. Find the fixed end moments and write it ready for analysis by Kani's method.

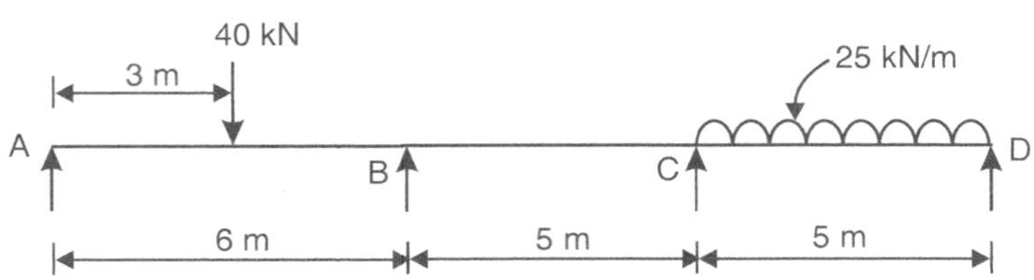

Fig. 9

Ans.

$$M_{FAB} = -\frac{Wl}{8} = -\frac{40 \times 6}{8} = -30 \text{ kNm}$$

$$M_{FBA} = +\frac{Wl}{8} = 30 \text{ kNm}$$

$$M_{FBC} = M_{FCB} = 0$$

$$M_{FCD} = -\frac{wl^2}{12} = -\frac{25 \times 5^2}{12} = -52.083 \text{ kNm}$$

$$M_{FDC} = +\frac{wl^2}{12} = +52.083 \text{ kNm}$$

For analysis by Kani's method, the moments at A and D are adjusted to account for the hinged ends at A and D as below :

$$M_{FAB} = 0$$

$$M_{FBA} = 30 + \frac{1}{2}(30) = 45 \text{ kNm}$$

$$M_{FCD} = -52.0893 - \frac{1}{2} \times 52.083 = -78.125 \text{ kNm} ; \quad M_{FDC} = 0$$

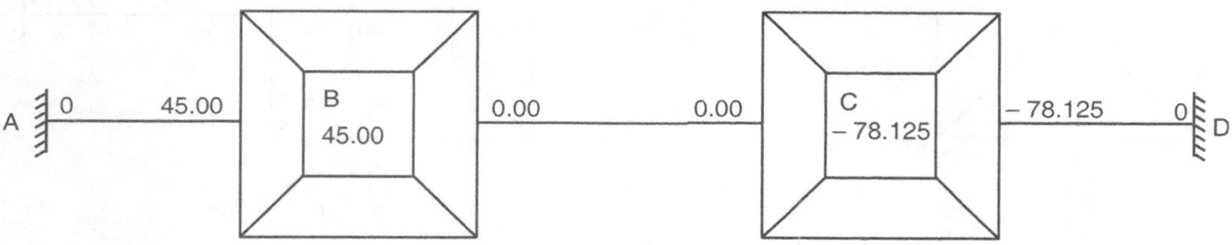

Fig. 9 (a)

12. For the continuous beam shown in Fig. 10, determine the rotation factors and show them in a figure.

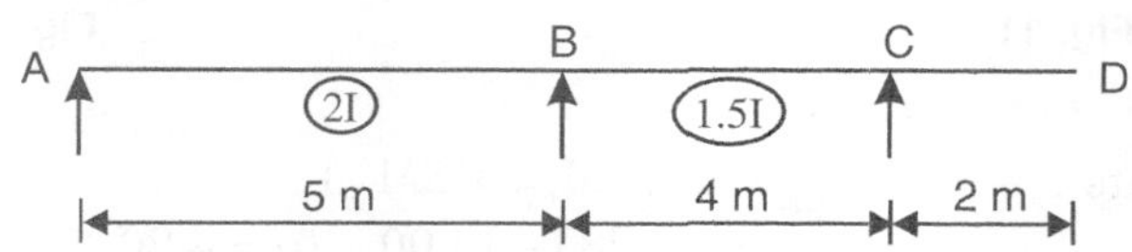

Fig. 10

Ans. Relative stiffness of BA $= \frac{3}{4} \times \frac{2I}{5} = \frac{3}{10} I$

Relative stiffness of BC $= \frac{3}{4} \times \frac{1.5I}{4} = \frac{4.5}{16} I$

$$\text{Sum} = \frac{3I}{10} + \frac{4.5I}{16} = \frac{24I + 22.5I}{80} = \frac{46.5I}{80}$$

Rotation factor u for BA $= -\frac{1}{2} \times \frac{3I}{10} \times \frac{80}{46.5I} = -0.258$

Rotation factor u for BC $= -\frac{1}{2} \times \frac{4.5I}{16} \times \frac{80}{46.5I} = -0.242$

Rotation factor u for CB $= -0.5$

Rotation factor u for CD $= 0$

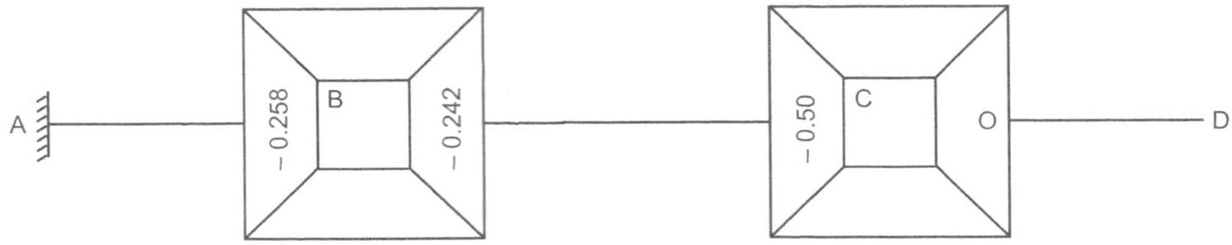

Fig. 10 (a) Rotation factors.

13. Carry out the first cycle of iteration in the frame shown in Fig. 11 using Kani's method. The fixed end moments due to external loads and rotation factors are shown.

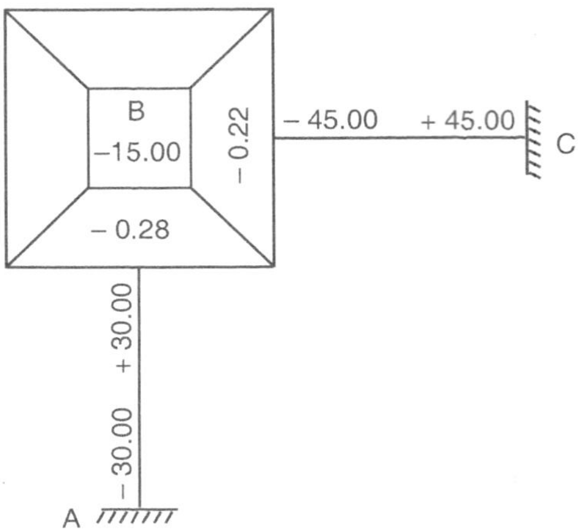

Fig. 11

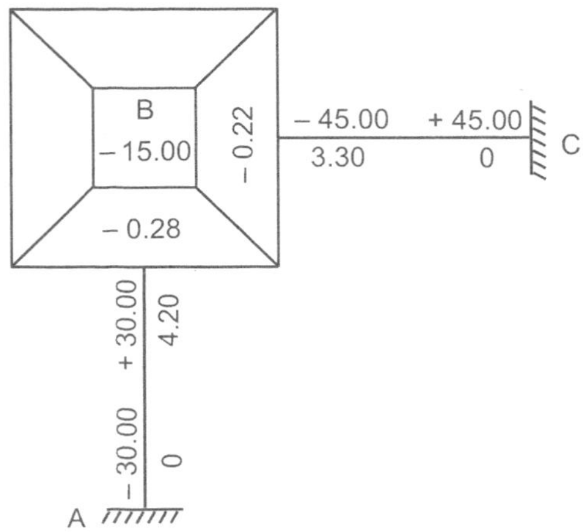

Fig. 11 (a)

Ans. I cycle :

Rotation moments :

$$M_{BA} = u_{BA} (M_{FB} + \Sigma M'_B)$$
$$= -0.28 (-15.00 + 0) = 4.20$$
$$M_{BC} = u_{BC} (M_{FB} + \Sigma M'_B)$$
$$= -0.22 (-15.00 + 0) = 3.30.$$

14. Calculate the displacement factors for the columns of the portal frame shown in Fig. 12.

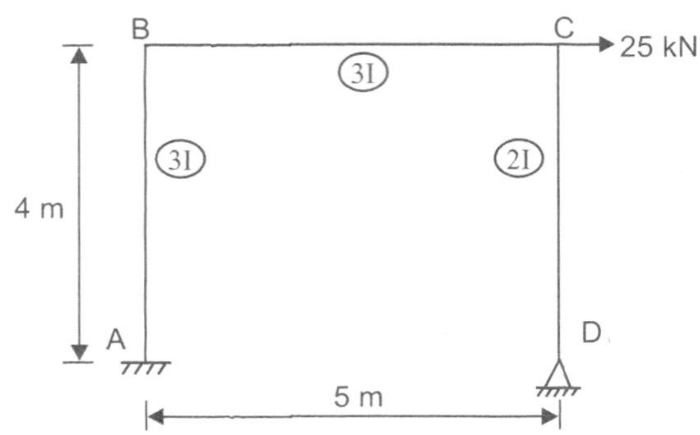

Fig. 12

Ans. For frames with equal height of columns, displacement factor $D_{ij} = -1.5 \times \dfrac{k_{ij}}{\Sigma(mk_{ij})}$

$m = 1$ for columns fixed at bottom.

$$D_{BA} = -1.5 \times \frac{k_{AB}}{k_{BA} + k_{CD}}$$

$$= -1.5 \times \frac{\left(\frac{2I}{4}\right)}{\left(\frac{2I}{4}\right) + \left(\frac{2I}{4}\right)} = -1.5 \times \frac{2I}{4} \times \frac{1}{I} = -\frac{3}{4} = -0.75$$

$$D_{CD} = -1.5 \times \frac{k_{CD}}{k_{BA} + k_{CD}}$$

$$= -1.5 \times \frac{\left(\frac{2I}{4}\right)}{\left(\frac{2I}{4}\right) + \left(\frac{2I}{4}\right)} = -1.5 \times \frac{2I}{4} \times \frac{1}{I} = -\frac{3}{4} = -0.75.$$

15. Calculate the displacement factors for the columns of the portal frame shown in Fig. 13.

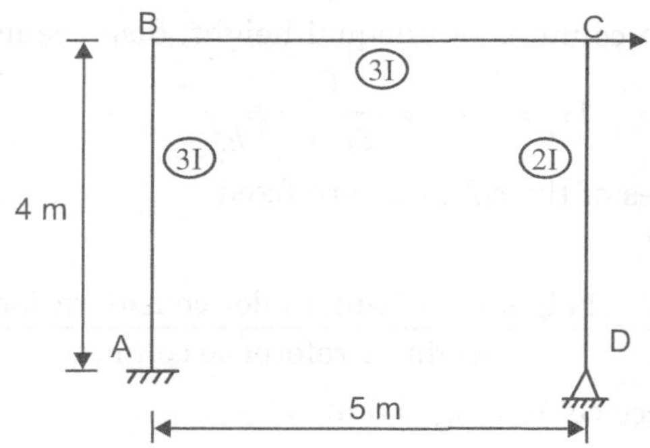

Fig. 13

Ans. Displacement factor, $D_{ij} = -1.5 \times \dfrac{k_{ij}}{\Sigma(mk_{ij})}$

$m = 1$ and $\dfrac{3}{4}$ for columns with fixed and hinged far end conditions respectively.

$$D_{BA} = -1.5 \times \frac{k_{BA}}{k_{BA} + k_{CD}}$$

$$= -1.5 \times \frac{\left(\frac{3I}{4}\right)}{\left(\frac{3I}{4}\right) + \left(\frac{3}{4} \times \frac{2I}{4}\right)} = -1.5 \times \left(\frac{3I}{4}\right) \times \frac{8}{9I} = -1.00$$

$$D_{CD} = -1.5 \times \frac{\left(\frac{3}{4} \times \frac{2I}{4}\right)}{\left(\frac{3I}{4}\right) + \left(\frac{3}{4} \times \frac{2I}{4}\right)} = -1.5 \times \frac{6I}{16} \times \frac{8}{9I} = -0.50.$$

16. Determine the displacement factors for the columns of the portal frame shown in Fig. 14.

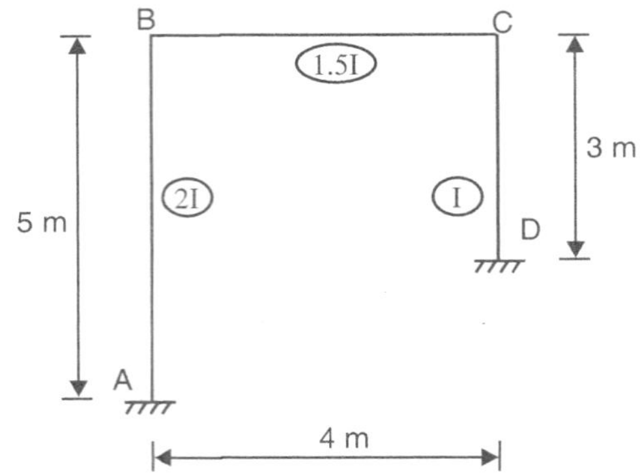

Fig. 14

Ans. For frames with columns of unequal height, displacement factor

$$D_{ij} = -1.5 \times \frac{C_{ij}\, k_{ij}}{\Sigma m\, C_{ij}^2\, k_{ij}}$$

$m = 1$ ($\because$ the bases of the columns are fixed)

C_{ij} = reduction factor

$$= \frac{h_i}{h_r} = \frac{\text{height of column under consideration}}{\text{height of reference column}}$$

Choosing AB as the reference column, $h_r = 5$ m

Column	C_{ij}	C_{ij}^2	k_{ij}	$C_{ij}\, k_{ij}$	$C_{ij}^2\, k_{ij}$	$D_{ij} = -1.5 \times \frac{C_{ij}\, k_{ij}}{\Sigma m C_{ij}^2\, k_{ij}}$
BA	$\frac{5}{5} = 1$	1	$\frac{2I}{5}$	$\frac{2I}{5}$	$\frac{2I}{5}$	$-1.5 \times \frac{2I}{5} \times \frac{5}{2.361} = -1.271$
CD	$\frac{3}{5} = 0.6$	0.36	$\frac{I}{3}$	$\frac{0.6I}{3} = \frac{I}{5}$	$\frac{0.36I}{5}$	$-1.5 \times \frac{I}{5} \times \frac{5}{2.361I} = -0.636$

$$\text{Sum} = \frac{2.361I}{5}$$

17. In Fig. 14, if columns AB and CD are of equal heights and both columns are hinged at the base, what will be the change in the formula for calculating displacement factors ?

Ans. $m = \frac{3}{4}$ and C_{ij} will be equal to 1. Hence $D_{ij} = -1.5 \times \frac{k_{ij}}{\Sigma k_{ij}}$.

18. Analysis by Kani's method has to be made on a symmetric frame symmetrically loaded as in Fig. 15. Sketch the simplified frame taking symmetry into account. I is the same for all members.

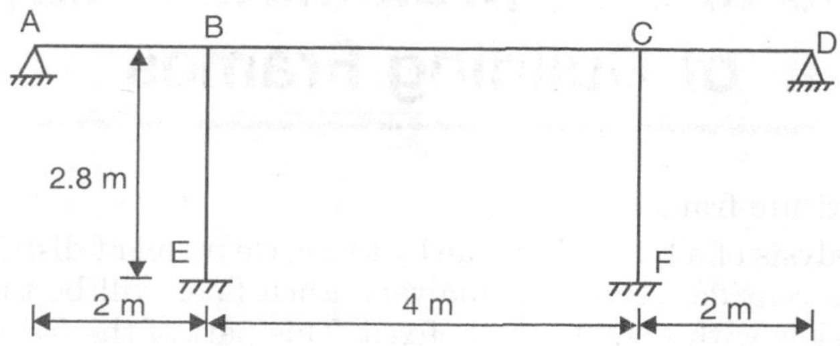

Fig. 15

Ans.

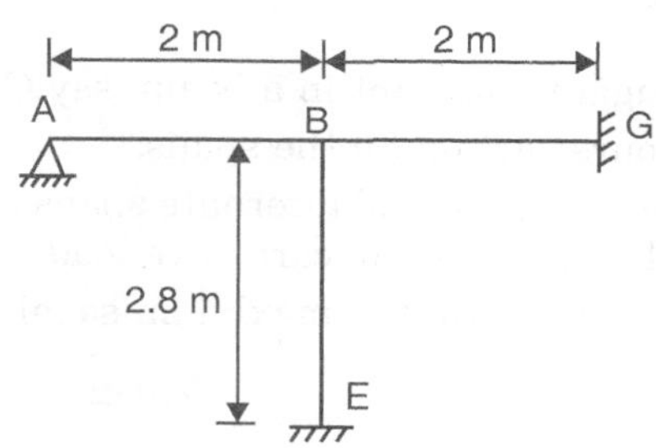

Fig. 15. (a) Simplified frame.

The simplified frame would be as shown in Fig. 15 (*a*).

19. Sketch the simplified frame if the frame in Fig. 15 were antisymmetrically loaded.
Ans. The simplified frame is shown in Fig. 16.

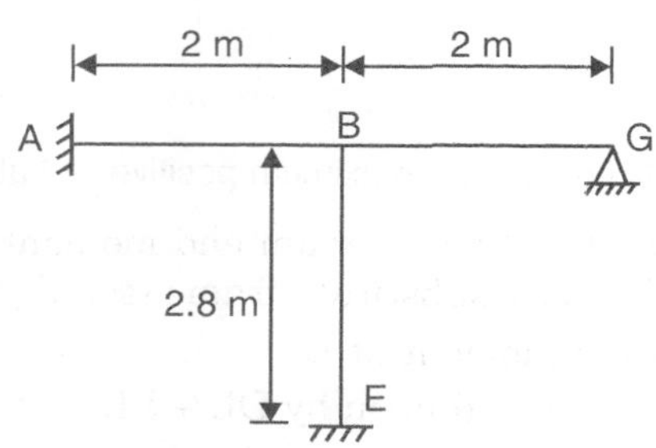

Fig. 16. Simplified frame.

22

Rapid and Approximate Analysis of Building Frames

1. What is a substitute frame ?

Ans. In the analysis of a building frame by two cycle moment distribution method, only a part of the frame is considered for the analysis. Each floor will be taken as connected to columns above and below with their far ends fixed. This part of the frame considered for the analysis is called a substitute frame.

2. With a neat sketch, illustrate the condition for maximum sagging moment in the beams of a substitute frame.

Ans. For maximum sagging moment in a beam, say CD in Fig. 1,

(*i*) dead load will of course act on all the spans.

(*ii*) the span under consideration and alternate spans (only two spans on each side need be considered) will be assumed to carry live load.

This loading condition will cause the maximum sagging moment in span CD.

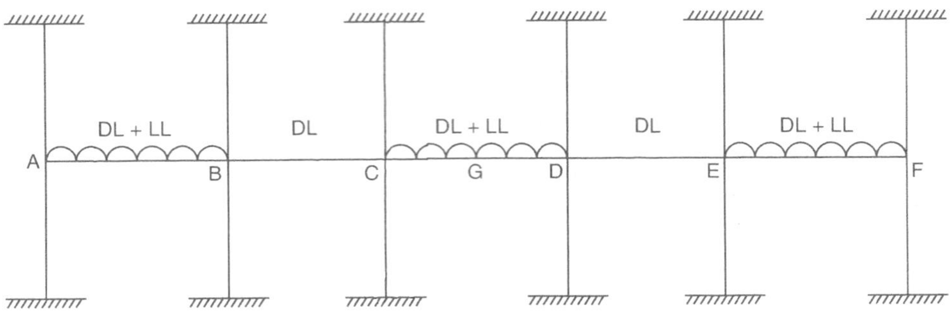

Fig. 1. Loading for maximum positive BM at G.

3. Give the loading condition for the maximum end moment in the left hand end support in the second span of a four bay substitute frame (see Fig. 2).

Ans. For maximum bending moment at B,

(*i*) Spans AB and BC will be acted upon by DL + LL

(*ii*) Span CD will be subjected to DL only and

(*iii*) Span DE need not be considered in this analysis (restricting to two spans to the right of B).

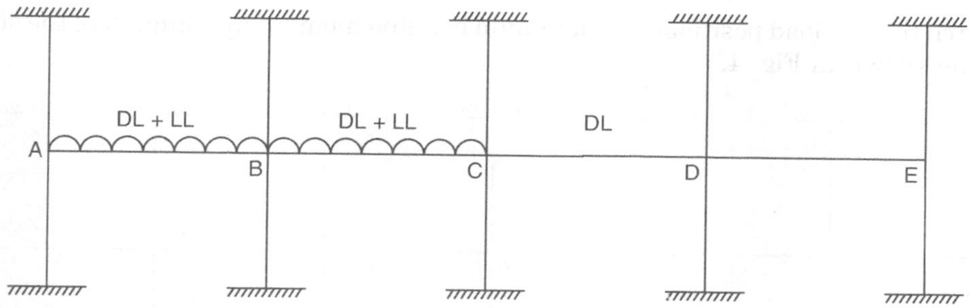

Fig. 2. Loading for maximum BM at B.

4. State the assumption made in the substitute frame analysis with regard to moment transfer from one floor to the other.

Ans. It is assumed that the moments transferred from one floor to the other are small and hence not considered. The moments for each floor are separately calculated.

5. In two cycle moment distribution method of analysis of a multi bay frame, how many bays are to be considered for the analysis of a particular bay/joint ?

Ans. Generally it is sufficient to consider two adjacent spans (if available) on each side of the bay/joint considered.

6. Give the loading condition for maximum negative bending moment at mid-span of the 4th bay of a seven bay substitute building frame.

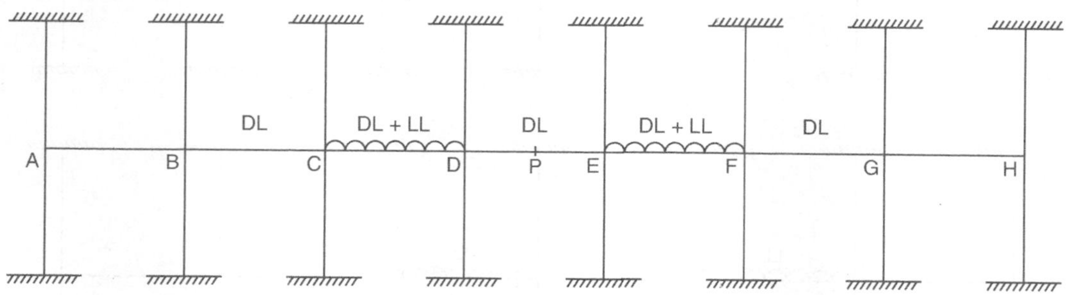

Fig. 3. Loading for maximum –ve BM at the mid-span of DE.

Ans. For maximum –ve B.M. at the mid-span of DE, *i.e.,* at P.

(*i*) The span DE will carry dead load only

(*ii*) Adjacent spans CD and EF will carry DL + LL

(*iii*) Spans beyond B on the left and beyond G on the right need not be considered in this analysis.

(*iv*) It is possible that under this loading, P may not develop any negative BM at all.

7. Mention the approximate methods of analysis of multistoreyed frames.

Ans. Approximate methods of analysis :

For vertical loads : Two cycle moment distribution method

For wind load (lateral load) : (1) Portal Method, (2) Cantilever Method, (3) Factor Method.

8. Sketch the live load positions for maximum bending moment in column A of the substitute frame shown in Fig. 4.

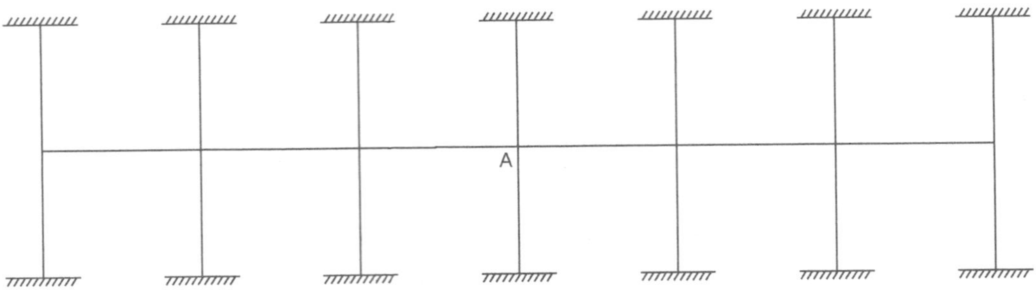

Fig. 4

Ans.

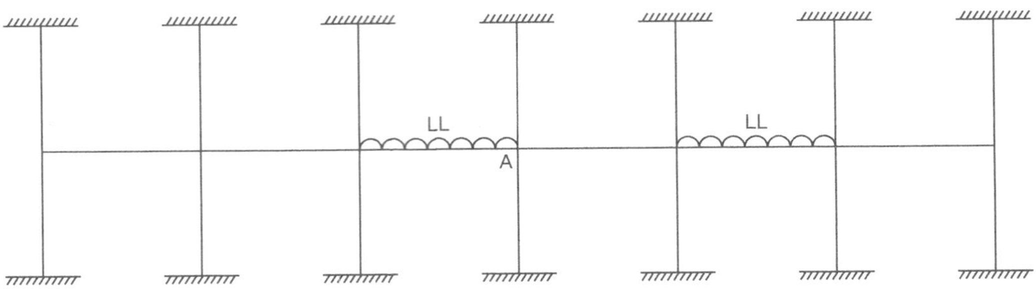

(a)

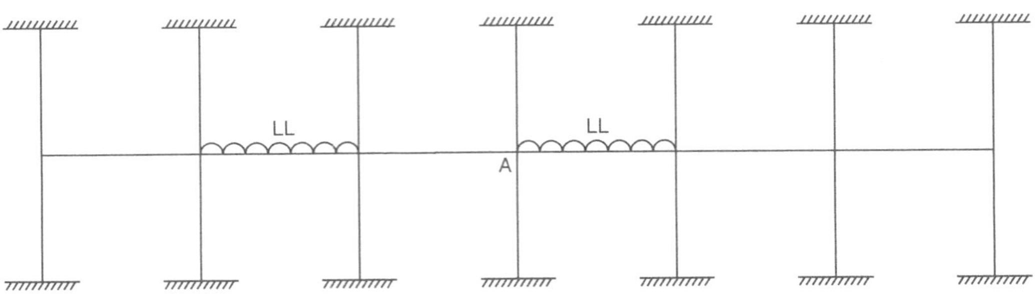

(b)

Fig. 5. (a), (b) Live load positions for maximum bending moment in column A.

9. Why do we prefer approximate methods for the analysis of building frames ?

Ans. Approximate methods are preferred because :

(*i*) Building frames being highly indeterminate, an exact analysis becomes tedious.

(*ii*) Approximate methods are also simple.

(*iii*) For practical building frame purposes, approximate methods can be considered sufficiently accurate, considering the fact that dead loads can easily vary by ± 10 per cent and live loads can never be estimated with any precision.

10. Using the portal method, determine the moments at the ends of the column FD of the portal frame shown in Fig. 6. Sketch the BMD.

Ans.

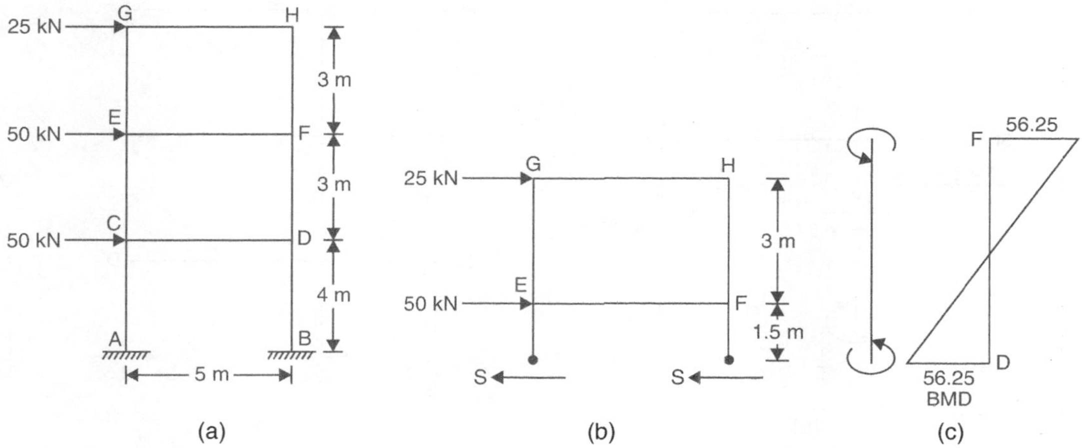

Fig. 6

$$\underset{(\leftarrow)}{S} + \underset{(\leftarrow)}{S} - \underset{(\rightarrow)}{25} - \underset{(\rightarrow)}{50} = 0$$

$$2S = 75$$

$$S = 37.5 \text{ kN}$$

Moment at F and D in DF $= 37.5 \times \frac{3}{2} = 56.25$ kN m (↻).

11. Using the portal method find the axial force in the columns BE and EH of the multistorey frame shown in Fig. 7, due to the lateral loads.

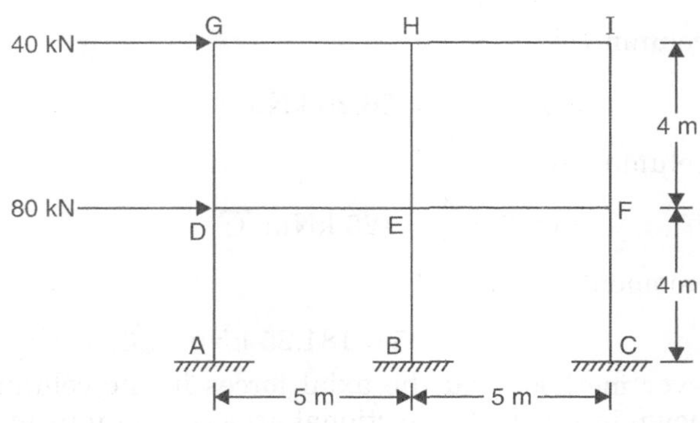

Fig. 7

Ans. Since the frame is symmetrical in geometry, axial force in the columns BE and EH due to the lateral loads will be ZERO.

12. Using the portal method, determine the moment at C of the beam CD in the frame shown in Fig. 6 (*a*) (Q. 10).

Ans.

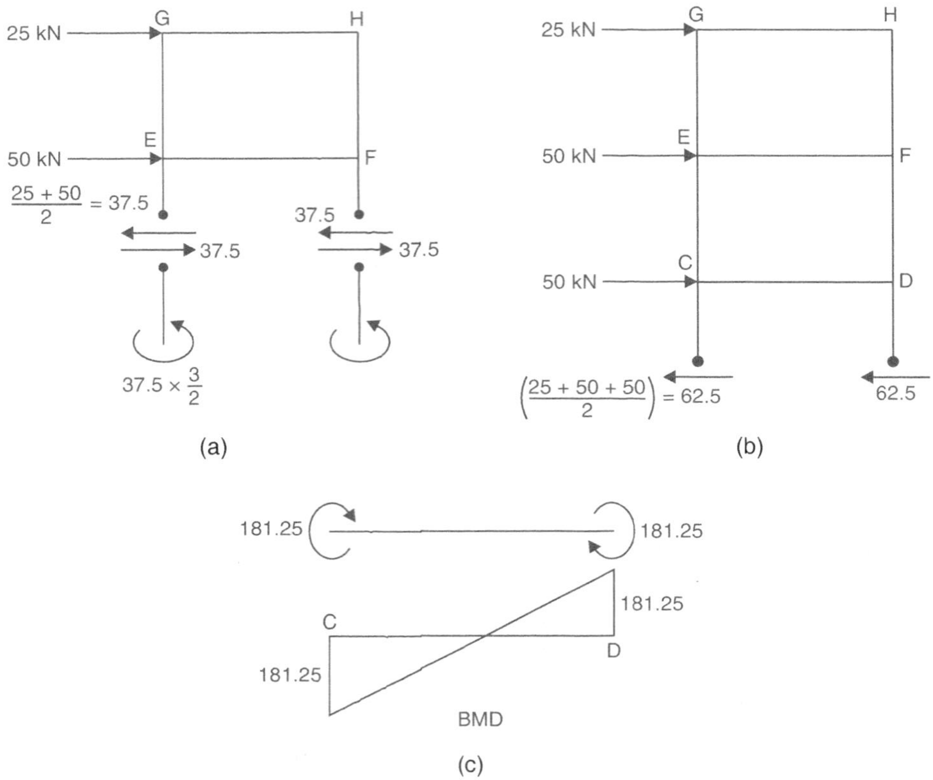

Fig. 8

Moment at C of column CE

$$= 37.5 \times \frac{3}{2} = 56.25 \text{ kNm } (\circlearrowleft)$$

Moment at C of column CA

$$= 62.5 \times \frac{4}{2} = 125 \text{ kNm } (\circlearrowleft)$$

For equilibrium, moment at C of CD

$$= 56.25 + 125 = 181.25 \text{ kNm } (\circlearrowright)$$

13. Using the cantilever method, find the axial forces in the columns AE and FB of the building frame shown in Fig. 9. The sectional areas of columns in m^2 are shown circled.

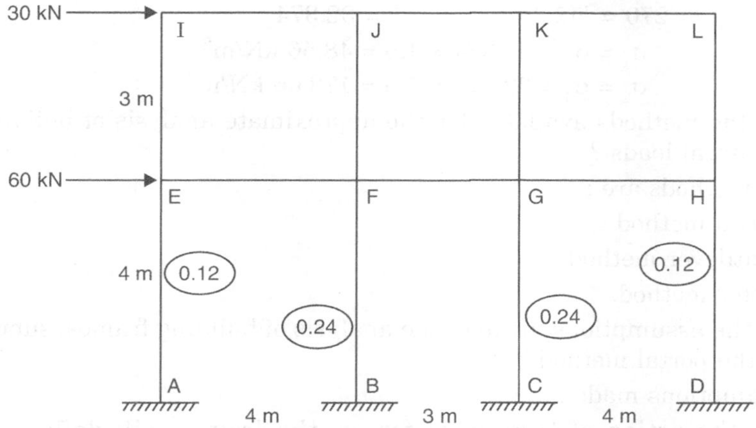

Fig. 9

Ans. Let us assume hinges at M, N, P and Q, mid-points of base columns (see Fig. 10). Over-turning moment about the line of hinges

$$M_1 = 30(3 + 2) + 60 \times 2$$
$$= 270 \text{ kNm}$$

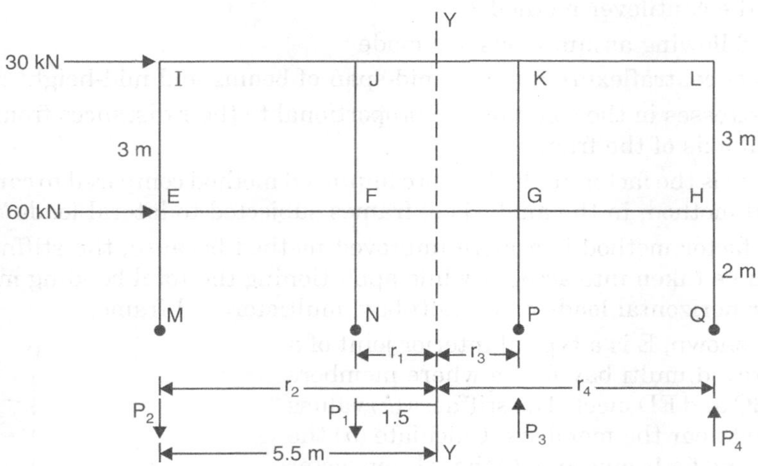

Fig. 10

Since the frame is symmetrical, the centroidal line YY, lies at 5.5 m from M and Q.

The r_i values are, $r_1 = r_3 = 1.5$ m ; $r_2 = r_4 = 5.5$ m.

The axial stresses can be assumed to be

$$\sigma_i = k.r_i$$

Hence moment due to any column $= kr_i\, a_i.r_i$

Hence, $M_1 = \Sigma k\, a_i r_1^2 = k \times 2(0.12 \times 5.5^2 + 0.24 \times 1.5^2)$

$$270 = k[8.34] \; ; \qquad k = 32.374$$
$$\sigma_1 = \sigma_3 = 32.374 \times 1.5 = 48.56 \text{ kN/m}^2$$
$$\sigma_2 = \sigma_4 = 32.374 \times 5.5 = 178.06 \text{ kN/m}^2.$$

14. What are the methods available for the approximate analysis of building frames subjected to lateral loads ?

Ans. The methods are :

1. The portal method
2. The cantilever method
3. The factor method.

15. What are the assumptions made in the analysis of building frames, subjected to lateral loads, by the portal method ?

Ans. Assumptions made :

1. Under the action of horizontal forces, the frames will deflect. The points of contraflexure occur at the middle points of all the members *i.e.,* beams and columns of the frame.
2. Frames having several bays are assumed to behave like a series of single bay portal frames placed side by side. Horizontal shear taken by each interior column, at the points of contraflexure, is thus double the horizontal shear taken by each of the external columns.

16. What are the assumptions made in the analysis of building frames, subjected to lateral loads, by the cantilever method ?

Ans. The following assumptions are made :

1. Points of contraflexure occur at mid-span of beams and mid-height of columns.
2. Axial stresses in the columns are proportional to their distances from the centroidal vertical axis of the frame.

17. In what way is the factor method a more improved method compared to cantilever method and portal method, in the analysis of frames subjected to lateral loads ?

Ans. The factor method is a more improved method because, the stiffnesses of beams and columns can be taken into account while apportioning the total bending moment induced by wind or other horizontal loads on a multi bay, multistoreyed frame.

18. In Fig. 11 shown, E is a typical interior joint of a multi-storeyed multi bay frame where members EA, EB, EC and ED meet. The stiffness (k) values are marked near the members. Calculate (a) the girder factor g for beams and (b) the column factor c of the columns.

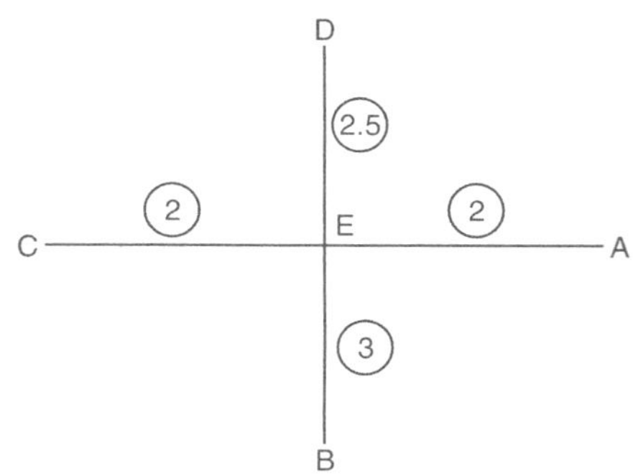

Fig. 11

Ans. Girder factor for beams at a joint :

$$g = \frac{\Sigma \, k \text{ of columns}}{\Sigma \, k \text{ of all members}}$$

$$g = \frac{2.5 + 3}{2.5 + 2 + 3 + 2} = \frac{5.5}{9.5} = 0.58$$

$\therefore$ Girder factor beams EC and EA at E are thus 0.58 and 0.58

Column factor c for each column at, joint E is given by

$$c = (1 - g)$$

$$c = 1 - 0.58 = 0.42$$

∴ Column factor for columns ED and EB are thus 0.42 and 0.42.

19. Obtain the moment factors for the members (C for columns and G for girders) in Fig. 11 (previous problem).

Ans. $c = 0.42$; $g = 0.58$ (as obtained earlier)

$$\text{Modified } c = 0.42 + \frac{0.42}{2} = 0.63 = c_m$$

$$\text{Modified } g = 0.58 + \frac{0.58}{2} = 0.87 = g_m$$

Moment factors are as follows:

$$\text{C for DE} = c_m \times k_{\text{DE}} = 0.63 \times 2.5 = 1.575$$

$$\text{C for BE} = c_m \times k_{\text{BE}} = 0.63 \times 3 = 1.89$$

$$\text{G for CE} = g_m \times k_{\text{CE}} = 0.87 \times 2 = 1.74$$

$$\text{G for AE} = g_m \times k_{\text{AE}} = 0.87 \times 2 = 1.74.$$

Bibliography

1. Punmia, B.C., Ashok Jain and Arun Jain, *Theory of Structures*, Laxmi Publications, New Delhi, 1992.
2. Gupta, S.P., Pandit, G.S., Gupta, R., *Theory of Structures*, Vol. I & II, Tata McGraw-Hill Publishing Company Limited, New Delhi, 1999.
3. Prakash Rao, D.S., *Structural Analysis*, University Press, 1996.
4. Ramamrutham, S., *Theory of Structures*, Dhanpat Rai & Sons, Delhi, 1986.
5. Reddy, C.S., *Basic Structural Analysis*, Tata McGraw-Hill Publishing Company Limited, New Delhi, 1997.
6. Khurmi, R.S., *Theory of Structures*, S. Chand & Company Ltd., New Delhi, 1998.
7. Junnarkar, S.B. and Shah, H.J., *Mechanics of Structures*—Volume II, Charotar Publishing House, Anand, 1996.
8. Negi, L.S. and Jangid, R.S., *Structural Analysis*, Tata McGraw-Hill Publishing Company Limited, New Delhi, 1997.
9. Anand, A.S., *Theory of Structures*, Satya Prakashan, New Delhi, 1989.
10. Vazirani, V.N., Ratwani, M.M., *Analysis of Structures*, Khanna Publishers, Delhi, 1976.
11. Ramchandra, Dr., *Design of Steel Structures*, Vol. II, Standard Book House, New Delhi, 1991.
12. Kinney, J.S., *Indeterminate Structural Analysis*, Addison-Wesley Publishing Company, Inc., Massachusetts, 1962.
13. Lecture Notes of the Authors.
14. Question papers from various universities.
15. Vaidyanathan, Dr. R. and Perumal, Dr. P., *Comprehensive Structural Analysis*, Volumes I and II, Laxmi Publications (P) Ltd., New Delhi-110002.

Index

L

M

N

P

R

S

T

U

V

W

Y